Linear Algebra

An Introduction

VNR NEW MATHEMATICS LIBRARY

under the general editorship of

J.V. Armitage
Principal
College of St. Hild and St. Bede
Durham University

N. Curle
Professor of Applied Mathematics
University of St. Andrews

The aim of this series is to provide a reliable modern coverage of those mainstream topics that form the core of mathematical instruction in universities and comparable institutions. Each book deals concisely with a well defined key area in pure or applied mathematics or statistics. Many of the volumes are intended not solely for students of mathematics, but also for engineering and science students whose training demands a firm grounding in mathematical methods.

Titles in the series:

Linear Algebra

An introduction

Second edition

A.O. Morris

Professor of Pure Mathematics
University College of Wales
Aberystwyth

CHAPMAN AND HALL

University and Professional Division

LONDON • NEW YORK • TOKYO • MELBOURNE • MADRAS

UK	Chapman and Hall, 11 New Fetter Lane, London EC4P 4EE
USA	Chapman and Hall, 29 West 35th Street, New York NY10001
JAPAN	Chapman and Hall Japan, Thomson Publishing Japan, Hirakawacho Nemoto Building, 7F, 1-7-11 Hirakawa-cho, Chiyoda-ku, Tokyo 102
AUSTRALIA	Chapman and Hall Australia, Thomas Nelson Australia, 480 La Trobe Street, PO Box 4725, Melbourne 3000
INDIA	Chapman and Hall India, R. Sheshadri, 32 Second Main Road, CIT East, Madras 600 035

First edition 1978
Reprinted 1978, 1980
Second edition 1982
Reprinted 1983, 1985, 1986, 1989, 1990

© 1978, 1982 A.O. Morris

Printed in Hong Kong by Thomas Nelson (Hong Kong) Ltd.

ISBN 0 412 38100 1

British Library Cataloguing in Publication Data
Morris, A.O. (Alun Owen), 1935–
    Linear algebra.

    (VNR new mathematics library; 9)
    Includes index.
    1. Algebras, Linear. I. Title.   II. Series.
QA184.M67   1983   512'.5   82-15890
ISBN 0 412 38100 1

Preface

This book is intended as an elementary introduction to linear algebra
and matrix theory for first year students in Universities and Polytechnics.
It is suitable for intending mathematicians, scientists, engineers and
social scientists who require a thorough grounding in linear algebra and
matrix theory as part of their course. It has been used in a class which
included both mathematicians and scientists and is based on a first year
course given at the University College of Wales, Aberystwyth over a
number of years.

Although there are many excellent books already available on this
subject, the author has not found one which is totally suitable for use
at this level. They tend in general to be too long and encyclopaedic and
the approach too abstract to be used as a course text in a first introduction to the subject. The approach used in this book emphasises the
computational and practical aspects of the subject, but at the same time
it aims to give a thorough and rigorous introduction to the subject at a
level which is suitable for use as a first introduction to abstract concepts
in mathematics. The book is structured so as to first give the more
concrete and practical aspects which lead naturally to the more abstract
ideas developed later.

The book's six chapters cover linear equations and matrices,
determinants, vector spaces, linear transformations on vector spaces,
inner product spaces and diagonalization of matrices and linear transformations. The first two chapters, after first motivating the discussion
by considering 2×2 and 3×3 matrices and determinants and the
corresponding linear equations, proceed to develop these concepts for
general $m \times n$ arrays. The main emphasis is in providing efficient and
effective techniques for solving linear equations, expanding determinants, manipulating matrices, etc. The later chapters are more abstract
in character, but each new abstract concept is motivated by considering
the relevant concept in 2- or 3-dimensions. The aim is to show that the
definitions of these abstract concepts are the natural and useful
extensions of what occurs in lower dimensions. Furthermore, it is
hoped that the student will develop an appreciation that abstraction

not only adds elegance to the subject but that it also leads quickly to effective techniques for dealing with certain practical problems. The exercises have also been selected to give the reader practical experience of the new concepts as they are introduced. They have only rarely been used to develop additional theory as is sometimes customary.

A rudimentary knowledge of algebra is assumed, e.g. complex numbers, fields, set theory, equivalence relations and mappings, which should be available to the reader through a concurrent or preceding course (or for some from their A-level courses). A general reference is P. J. Higgins—A First Course in Abstract Algebra, VNR New Mathematics Library 7.

The book has benefited from the detailed reading of my colleague Mr. Meurig John; I am grateful to him for his ready assistance as I am to Mrs. Noreen Davies who patiently typed the manuscript.

The University College of Wales   Alun O. Morris
Aberystwyth
March 1977

Preface to the Second Edition

The second edition of this book is largely a reproduction of the first edition. I have taken advantage of the opportunity to clarify a few obscurities and to correct some minor errors in the first edition. I am indebted to my colleagues who have drawn my attention to these.

The major changes have resulted in more examples of vector spaces and linear transformations with an analytic flavour, a complete rewriting of the effect of a change of basis on the matrix of a linear transformation and a fuller treatment on the application of quadratic forms to the classification of conic sections and quadric surfaces. At the suggestion of many readers, solutions for all the exercises have been included.

The University College of Wales   Alun O. Morris
Aberystwyth
June 1982

Contents

CHAPTER 1

Linear Equations and Matrices

1.1 Introduction

The reader is probably already familiar with some methods for solving systems of linear equations where a small number of equations and variables are involved. Applications of linear algebra to other branches of science, engineering, economics or elsewhere generally occur via the need to solve such systems of linear equations. It may be claimed that one of the main aims of linear algebra is to

(i) find the most economic way of manipulating and solving such systems,

(ii) obtain useful theoretical results concerning such systems.

At an elementary level, the method normally used for the solution of a system of linear equations is one involving either the use of determinants or the use of an elimination process. In this chapter, our intention is to develop this second method to the general case. This turns out to be the most efficient method of dealing with such a system.

As motivation for the work to be developed in the remainder of this chapter, we first illustrate the work by considering a few examples.

EXAMPLE   Find all the solutions of the following systems of linear equations:

(i) $\quad x_1 - 2x_2 + x_3 = 1$
$\qquad 2x_1 - x_2 + x_3 = 2$
$\qquad 4x_1 + x_2 - x_3 = 1$

(ii) $\quad x_1 + x_2 = 2$
$\qquad 2x_1 + 2x_2 = 3$

(iii) $\quad x_1 - 2x_2 + x_3 = 1$
$\qquad 2x_1 - x_2 + x_3 = 2$

(i) Subtract twice and four times the first equation from the second and third equations respectively, giving the system of equations

$$x_1 - 2x_2 + x_3 = 1$$
$$3x_2 - x_3 = 0$$
$$9x_2 - 5x_3 = -3$$

1

Now, divide the second equation by 3 and add twice the resulting equation to the first equation and subtract nine times the resulting equation from the third equation, giving

$$x_1 \quad + \tfrac{1}{3}x_3 = 1$$
$$x_2 - \tfrac{1}{3}x_3 = 0$$
$$- 2x_3 = -3$$

which leads to the solution

$$x_1 = -\tfrac{1}{2} \quad x_2 = \tfrac{1}{2} \quad x_3 = \tfrac{3}{2}$$

this is the unique solution of the original system of equations. What we have done is to replace in a systematic way the original system of equations by a system of equations "equivalent" to it whose solution is more easily obtained.

(ii) These two equations are inconsistent, since subtracting twice one equation from the other leads to the fallacious statement $1 = 0$, thus no solution exists in this case.

(iii) The system of equations may be reduced to the system of equations

$$x_1 \quad + \tfrac{1}{3}x_3 = 1$$
$$x_2 - \tfrac{1}{3}x_3 = 0$$

or in other words

$$x_1 = 1 - \tfrac{1}{3}x_3$$
$$x_2 = \tfrac{1}{3}x_3$$

If we now put $x_3 = t$, we find that whatever value we assign to t, we obtain a solution of the original system of equations. For example, if $t = 0, x_1 = 1, x_2 = x_3 = 0$ is a solution and if $t = 3, x_1 = 0, x_2 = 1$, $x_3 = 3$ is another solution. In this case, not only does a solution exist, but there is more than one solution to the system.

From the above examples, we note that a system may have
(i) no solution, (ii) a unique solution, or (iii) more than one solution. Furthermore, the process of solution is to use a systematic method of elimination which leads to a "simpler equivalent" system of equations whose solution is more easily computed.

Initially, we shall be most interested in the technique of solution of linear equations, later we shall take up further the questions
(i) When do solutions exist?
(ii) When is a solution the unique solution?

2

1.2 Elementary Row Operations on Matrices

Let K be a field.

DEFINITION 1.1   *A rectangular array of elements of K*

$$\begin{pmatrix} \alpha_{11} & \alpha_{12} & \ldots & \alpha_{1n} \\ \alpha_{21} & \alpha_{22} & \ldots & \alpha_{2n} \\ . & . & & . \\ . & . & & . \\ . & . & & . \\ \alpha_{m1} & \alpha_{m2} & \ldots & \alpha_{mn} \end{pmatrix}$$

is called a **matrix**. *For $i = 1, 2, \ldots, m$, let*

$$r_i = (\alpha_{i1}, \alpha_{i2}, \ldots, \alpha_{in})$$

and for $j = 1, 2, \ldots, n$, let

$$c_j = \begin{pmatrix} \alpha_{1j} \\ \alpha_{2j} \\ . \\ . \\ . \\ \alpha_{mj} \end{pmatrix}$$

then the r_i ($i = 1, \ldots, m$) are called the **rows** *of the matrix and the c_j ($j = 1, \ldots, n$) are called the* **columns** *of the matrix. A matrix with m rows and n columns is called an* **$m \times n$ matrix**. *The element of K at the intersection of the ith row and jth column is called the (i, j)th* **element** *of the matrix. The matrix is written in abbreviated form as*

$$A = (\alpha_{ij})_{m \times n} \quad \text{or} \quad A = (\alpha_{ij})$$

EXAMPLE

$$\begin{pmatrix} 1 & 0 & -3 & 1 \\ 2 & 1 & 3 & 1 \\ 1 & 0 & 1 & 1 \end{pmatrix}$$

is a 3×4 matrix, where $(2, 1, 3, 1)$ is a row of the matrix and $\begin{pmatrix} -3 \\ 3 \\ 1 \end{pmatrix}$ is a column of the matrix.

DEFINITION 1.2   *The following are called* **elementary row operations** *on a matrix $A = (\alpha_{ij})$,*

3

(i) *if $1 \le i \le m$, $\alpha \in K$, $\alpha \ne 0$, multiply the ith row of A by α,*
(ii) *if $1 \le i, j \le m$, $i \ne j$, $\alpha \in K$, add α times the jth row to the ith row,*
(iii) *interchange the ith and jth rows.*

If r_i $(i = 1, \ldots, m)$ are the m rows of A, these three elementary row operations are written as follows

(i) $r_i \to \alpha r_i$, (ii) $r_i \to r_i + \alpha r_j$, (iii) $r_i \leftrightarrow r_j$.

DEFINITION 1.3 *If A and B are two $m \times n$ matrices, then A is **row equivalent** to B if B can be obtained from A by a finite sequence of elementary row operations.*

EXAMPLE

$$\begin{pmatrix} 1 & 0 & -1 & 1 \\ 2 & 1 & 0 & 1 \\ -1 & 1 & 0 & 2 \end{pmatrix}$$

is row equivalent to

$$\begin{pmatrix} 1 & 0 & -1 & 1 \\ -1 & 1 & 0 & 2 \\ 0 & 3 & 0 & 5 \end{pmatrix}$$

since

$$\begin{pmatrix} 1 & 0 & -1 & 1 \\ 2 & 1 & 0 & 1 \\ -1 & 1 & 0 & 2 \end{pmatrix} \xrightarrow{r_2 \to r_2 + 2r_3} \begin{pmatrix} 1 & 0 & -1 & 1 \\ 0 & 3 & 0 & 5 \\ -1 & 1 & 0 & 2 \end{pmatrix}$$

$$\xrightarrow{r_2 \leftrightarrow r_3} \begin{pmatrix} 1 & 0 & -1 & 1 \\ -1 & 1 & 0 & 2 \\ 0 & 3 & 0 & 5 \end{pmatrix}$$

DEFINITION 1.4 *An $m \times n$ matrix A is called an **echelon matrix** if*
(i) *the first non-zero element in each non-zero row is 1,*
(ii) *the leading 1 in any non-zero row occurs to the right of the leading 1 in any preceding row,*
(iii) *the non-zero rows appear before the zero rows.*
*An echelon matrix is called a **reduced echelon matrix** if*
(iv) *the leading 1 in any non-zero row is the only non-zero element in the column in which that 1 occurs.*

EXAMPLE

$$\begin{pmatrix} 1 & 0 & -1 & 2 \\ 0 & 1 & 1 & 3 \\ 0 & 0 & 1 & 1 \end{pmatrix}$$

is an echelon matrix, but not a reduced echelon matrix,

$$\begin{pmatrix} 1 & 0 & 0 & 3 \\ 0 & 1 & 0 & 2 \\ 0 & 0 & 1 & 1 \end{pmatrix}$$

is a reduced echelon matrix, but

$$\begin{pmatrix} 0 & 1 & 0 & 2 \\ 1 & 0 & 2 & 0 \\ 0 & 0 & 0 & 0 \end{pmatrix} \quad \text{and} \quad \begin{pmatrix} 1 & 0 & 2 & 0 \\ 0 & 0 & 0 & 0 \\ 0 & 1 & 0 & 2 \end{pmatrix}$$

are not even echelon matrices.

We can now prove

THEOREM 1.5   *Every $m \times n$ matrix is row equivalent to an $m \times n$ reduced echelon matrix.*

PROOF   (*Gauss Elimination Method*) Let the jth column of A, where $1 \leqslant j \leqslant n$, be the first column of A which contains a non-zero element. Suppose that $\alpha_{ij} \neq 0$, where $1 \leqslant i \leqslant m$, then the elementary row operation $r_i \to \dfrac{1}{\alpha_{ij}} r_i$ followed by the elementary row operation $r_1 \leftrightarrow r_i$ gives a matrix with 1 in the $(1,j)$-position, viz:

$$\begin{pmatrix} 0 & \cdots & 0 & 1 & \beta_{1,j+1} & \cdots & \beta_{1n} \\ 0 & \cdots & 0 & \beta_{2j} & \beta_{2,j+1} & \cdots & \beta_{2n} \\ \vdots & & \vdots & & & & \vdots \\ 0 & \cdots & 0 & \beta_{mj} & \beta_{m,j+1} & \cdots & \beta_{mn} \end{pmatrix}$$

Now, the elementary row operations $r_k \to r_k - \beta_{kj} r_1$ $(k = 2, \ldots, m)$ result in the matrix

$$\begin{pmatrix} 0 & \cdots & 0 & 1 & \gamma_{1,\,j+1} & \cdots & \gamma_{1n} \\ 0 & \cdots & 0 & 0 & \gamma_{2,\,j+1} & \cdots & \gamma_{2n} \\ \cdot & & \cdot & \cdot & & & \cdot \\ \cdot & & \cdot & \cdot & & & \cdot \\ 0 & \cdots & 0 & 0 & \gamma_{m,\,j+1} & \cdots & \gamma_{mn} \end{pmatrix}$$

If at this stage, some zero rows have appeared, row interchanges can be used so that the non-zero rows appear before the zero rows. We now repeat the same process with the last $m-1$ rows of this matrix to form

$$\begin{pmatrix} 0 & \cdots & 0 & 1 & \delta_{1,\,j+1} & \cdots & \delta_{1,\,k-1} & \delta_{1k} & \delta_{1,\,k+1} & \cdots & \delta_{1n} \\ 0 & \cdots & 0 & 0 & 0 & \cdots & 0 & 1 & \delta_{2,\,k+1} & \cdots & \delta_{2n} \\ 0 & \cdots & 0 & 0 & 0 & \cdots & 0 & \delta_{3k} & \delta_{3,\,k+1} & \cdots & \delta_{3n} \\ \cdot & & \cdot & \cdot & \cdot & & \cdot & \cdot & \cdot & & \cdot \\ \cdot & & \cdot & \cdot & \cdot & & \cdot & \cdot & \cdot & & \cdot \\ 0 & \cdots & 0 & 0 & 0 & \cdots & 0 & \delta_{mk} & \delta_{m,\,k+1} & \cdots & \delta_{mn} \end{pmatrix} \quad (1)$$

After a finite number of steps a matrix which is in echelon form results. A matrix which is in reduced echelon form is obtained from this by carrying out a sequence of elementary row operations of the type 1.2 (ii), for example, the sequence of elementary row operations $r_i \rightarrow r_i - \delta_{ik}\,r_2\ (i=1,3,\ldots,m)$ reduces the matrix (1) to the form

$$\begin{pmatrix} 0 & \cdots & 0 & 1 & \epsilon_{1,\,j+1} & \cdots & \epsilon_{1,\,k-1} & 0 & \epsilon_{1,\,k+1} & \cdots & \epsilon_{1n} \\ 0 & \cdots & 0 & 0 & 0 & \cdots & 0 & 1 & \epsilon_{2,\,k+1} & \cdots & \epsilon_{2n} \\ 0 & \cdots & 0 & 0 & 0 & \cdots & 0 & 0 & \epsilon_{3,\,k+1} & \cdots & \epsilon_{3n} \\ \cdot & & \cdot & \cdot & \cdot & & \cdot & \cdot & \cdot & & \cdot \\ \cdot & & \cdot & \cdot & \cdot & & \cdot & \cdot & \cdot & & \cdot \\ 0 & \cdots & 0 & 0 & 0 & \cdots & 0 & 0 & \epsilon_{m,\,k+1} & \cdots & \epsilon_{mn} \end{pmatrix}$$

This completes the proof of the theorem. ∎

The above proof will be better understood if the following example is worked out in detail.

EXAMPLE

Let

$$A = \begin{pmatrix} 0 & 0 & 5 & 35 & -24 & 1 \\ 0 & 2 & 1 & -1 & 1 & 0 \\ 0 & 3 & 2 & 2 & -1 & 1 \\ 0 & 0 & 0 & 0 & 0 & 0 \\ 0 & 5 & 3 & 1 & 0 & 1 \end{pmatrix} \quad (2)$$

6

The elementary row operations $r_2 \to \frac{1}{2}r_2$, $r_1 \leftrightarrow r_2$ and $r_4 \leftrightarrow r_5$ gives the matrix

$$\begin{pmatrix} 0 & 1 & \frac{1}{2} & -\frac{1}{2} & \frac{1}{2} & 0 \\ 0 & 0 & 5 & 35 & -24 & 1 \\ 0 & 3 & 2 & 2 & -1 & 1 \\ 0 & 5 & 3 & 1 & 0 & 1 \\ 0 & 0 & 0 & 0 & 0 & 0 \end{pmatrix}$$

The elementary row operations $r_3 \to r_3 - 3r_1$, $r_4 \to r_4 - 5r_1$ give the matrix

$$\begin{pmatrix} 0 & 1 & \frac{1}{2} & -\frac{1}{2} & \frac{1}{2} & 0 \\ 0 & 0 & 5 & 35 & -24 & 1 \\ 0 & 0 & \frac{1}{2} & \frac{7}{2} & -\frac{5}{2} & 1 \\ 0 & 0 & \frac{1}{2} & \frac{7}{2} & -\frac{5}{2} & 1 \\ 0 & 0 & 0 & 0 & 0 & 0 \end{pmatrix}$$

Now, the elementary row operations $r_2 \leftrightarrow r_3$, $r_1 \to r_1 - r_2$, $r_4 \to r_4 - r_2$, $r_3 \to r_3 - 10r_2$, $r_2 \to 2r_2$ performed in that order give the matrix

$$\begin{pmatrix} 0 & 1 & 0 & -4 & 3 & -1 \\ 0 & 0 & 1 & 7 & -5 & 2 \\ 0 & 0 & 0 & 0 & 1 & -9 \\ 0 & 0 & 0 & 0 & 0 & 0 \\ 0 & 0 & 0 & 0 & 0 & 0 \end{pmatrix}$$

and finally, $r_1 \to r_1 - 3r_3$, $r_2 \to r_2 + 5r_3$ now gives the reduced echelon matrix

$$\begin{pmatrix} 0 & 1 & 0 & -4 & 0 & 26 \\ 0 & 0 & 1 & 7 & 0 & -43 \\ 0 & 0 & 0 & 0 & 1 & -9 \\ 0 & 0 & 0 & 0 & 0 & 0 \\ 0 & 0 & 0 & 0 & 0 & 0 \end{pmatrix} \qquad (3)$$

Note: It is important that these elementary row operations should be carried out in the order stated.

Exercises 1.2

1. Find the reduced echelon matrices of the following matrices

(i) $\begin{pmatrix} 1 & -1 & 2 & 1 \\ 2 & 1 & -1 & 1 \\ 1 & -2 & 1 & 1 \end{pmatrix}$

(ii) $\begin{pmatrix} 1 & 1 & -1 \\ 1 & -1 & 2 \\ 2 & 0 & 2 \\ 2 & 1 & -1 \end{pmatrix}$

(iii) $\begin{pmatrix} 2 & 2 & 5 & 3 \\ 6 & 1 & 5 & 4 \\ 4 & -1 & 0 & 1 \\ 2 & 0 & 1 & 1 \end{pmatrix}$

(iv) $\begin{pmatrix} 0 & 1 & 1 & 1 & 2 & 2 \\ -1 & 4 & 3 & 3 & 4 & 7 \\ 2 & 1 & 3 & 2 & 8 & 3 \\ 3 & 1 & 4 & -1 & 4 & 0 \\ 5 & 2 & 7 & 0 & 10 & 2 \end{pmatrix}$

(v) $\begin{pmatrix} i & 1-i & i & 0 \\ 1 & -2 & 0 & i \\ 1-i & -1+i & 1 & 1 \end{pmatrix}$

(vi) $\begin{pmatrix} 1 & 1-\sqrt{2} & 0 & \sqrt{2} \\ \sqrt{2} & -3 & 1+\sqrt{2} & -1-2\sqrt{2} \\ -1 & \sqrt{2} & -1 & 1 \\ \sqrt{2}-2 & -2+4\sqrt{2} & -2-\sqrt{2} & 3+\sqrt{2} \end{pmatrix}$

2. Are the following pairs of matrices row equivalent?

(i) $\begin{pmatrix} 1 & 0 & -1 \\ 2 & 1 & 0 \\ 1 & -1 & 1 \end{pmatrix}$ $\begin{pmatrix} 3 & -1 & 1 \\ 0 & 2 & 1 \\ 1 & -1 & 1 \end{pmatrix}$

(ii) $\begin{pmatrix} 1 & -1 & 1 & 2 \\ -2 & 3 & 0 & 1 \\ 1 & 0 & -1 & 3 \end{pmatrix}$ $\begin{pmatrix} 0 & -1 & 2 & 3 \\ 1 & 2 & -1 & 0 \\ -2 & -5 & 4 & 3 \end{pmatrix}$

8

1.3 Application to Linear Equations

Consider the system of m linear equations in n variables $x_1, \ldots, x_n$,

$$\sum_{j=1}^{n} \alpha_{ij} x_j = \beta_i \qquad (i = 1, \ldots, m) \tag{4}$$

where $\alpha_{ij}, \beta_i \in K$, then

$$A = (\alpha_{ij})_{m \times n} \quad \text{and} \quad (A|b) = (\alpha_{ij}|\beta_i)_{m \times (n+1)}$$

are called the **matrix of coefficients** and the **augmented matrix** of the system respectively, i.e.

$$A = \begin{pmatrix} \alpha_{11} & \alpha_{12} & \cdots & \alpha_{1n} \\ \alpha_{21} & \alpha_{22} & \cdots & \alpha_{2n} \\ \cdot & \cdot & & \cdot \\ \cdot & \cdot & & \cdot \\ \cdot & \cdot & & \cdot \\ \alpha_{m1} & \alpha_{m2} & \cdots & \alpha_{mn} \end{pmatrix} \quad \text{and}$$

$$(A|b) = \begin{pmatrix} \alpha_{11} & \alpha_{12} & \cdots & \alpha_{1n} & \beta_1 \\ \alpha_{21} & \alpha_{22} & \cdots & \alpha_{2n} & \beta_2 \\ \cdot & \cdot & & \cdot & \cdot \\ \cdot & \cdot & & \cdot & \cdot \\ \cdot & \cdot & & \cdot & \cdot \\ \alpha_{m1} & \alpha_{m2} & \cdots & \alpha_{mn} & \beta_m \end{pmatrix}$$

DEFINITION 1.6   *An n-tuple $(x_1, \ldots, x_n)$ which satisfies each of the m equations in the system* (4) *is called a* **solution** *of the system. Two systems of linear equations are* **equivalent** *if every solution of one system is a solution of the other system and vice versa. A system of linear equations is called a* **homogeneous system** *if $\beta_i = 0$ $(i = 1, \ldots, m)$. A system with at least one solution is called a* **consistent** *system. The solution $(0, \ldots, 0)$ of a homogeneous system is called the* **trivial** *solution.*

Our main result is the following:

THEOREM 1.7   *If $(A'|b') = (\alpha_{ij}'|\beta_i')$ is an $m \times (n + 1)$ matrix obtained from the $m \times (n + 1)$ matrix $(A|b)$ by an elementary row operation, then the systems $\sum_{j=1}^{n} \alpha_{ij} x_j = \beta_i$ $(i = 1, \ldots, m)$ and $\sum_{j=1}^{n} \alpha_{ij}' x_j = \beta_i'$ $(i = 1, \ldots, m)$ are equivalent.*

9

PROOF   We first note that if a matrix $(A'|b')$ is obtained from a matrix $(A|b)$ by an elementary row operation, then there exists an elementary row operation on $(A'|b')$ which results in the matrix $(A|b)$, this is the inverse of the original elementary row operation. The inverses of the three elementary row operations

(i) $r_i \rightarrow \alpha r_i, \alpha \neq 0$ (ii) $r_i \rightarrow r_i + \alpha r_j$ and   (iii) $r_i \leftrightarrow r_j$

are

(i)' $r_i \rightarrow \alpha^{-1} r_i$ (ii)' $r_i \rightarrow r_i - \alpha r_j$ and   (iii)' $r_i \leftrightarrow r_j$

respectively. Also, whatever elementary row operation has been used to obtain the matrix $(A'|b')$ from $(A|b)$, the resulting corresponding system of linear equations is a linear combination of the equations in the original system, and so, a solution of the original system will also be a solution of the new system. For example, for an elementary row operation of type (ii), the only equation changed is the ith equation, where the equation

$$\alpha_{i1} x_1 + \ldots + \alpha_{in} x_n = \beta_i$$

is replaced by

$$(\alpha_{i1} + \alpha \alpha_{j1}) x_1 + \ldots + (\alpha_{in} + \alpha \alpha_{jn}) x_n = \beta_i + \alpha \beta_j$$

and thus, if $(x_1, \ldots, x_n)$ is a solution of the original system, it is also a solution of the new system. Conversely, by the above, each solution of the new system can similarly be shown to be a solution of the original system. ∎

There are some important consequences of this theorem which appear in the following corollaries.

COROLLARY 1   *If $\sum\limits_{j=1}^{n} \alpha_{ij} x_j = \beta_i (i = 1, \ldots, m)$ is a system of linear equations with augmented matrix $(A|b) = (\alpha_{ij}|\beta_i)$ and $(R|s) = (\rho_{ij}|\sigma_i)$ is its reduced echelon matrix, then $\sum\limits_{j=1}^{n} \rho_{ij} x_j = \sigma_i (i = 1, \ldots, m)$ is equivalent to $\sum\limits_{j=1}^{n} \alpha_{ij} x_j = \beta_i (i = 1, \ldots, m)$.*

PROOF   By Theorem 1.5 the matrix $(R|s)$ is obtained from the matrix $(A|b)$ by a finite sequence of elementary row operations. The result now follows directly from the above theorem. ∎

It is useful to look at this result in more detail. Suppose that in the reduced echelon matrix $(R|s)$, the leading elements appear in columns

$j_1, j_2, \ldots, j_\ell$, and that the remaining columns are $j_{\ell+1}, \ldots, j_n$, then

$$x_{j_1} + \sum_{k=\ell+1}^{n} \rho_{1j_k} x_{j_k} = \sigma_1$$

$$x_{j_2} + \sum_{k=\ell+1}^{n} \rho_{2j_k} x_{j_k} = \sigma_2$$

$$x_{j_\ell} + \sum_{k=\ell+1}^{n} \rho_{\ell j_k} x_{j_k} = \sigma_\ell$$

$$0 = \sigma_{\ell+1}$$

where $\ell + 1 \leqslant m$. Then, either

(i) $\sigma_{\ell+1} = 1$ and $\sigma_1 = \ldots = \sigma_\ell = 0$, in which case the system is *not* consistent

or

(ii) $\sigma_{\ell+1} \equiv 0$ and the system *is* consistent, i.e. the system has a solution and this solution may now be written out explicitly by giving arbitrary values to $x_{j_{\ell+1}}, \ldots, x_{j_n}$.

If $\ell = n$, then the system has the unique solution $x_1 = \sigma_1, \ldots,$ $x_n = \sigma_n$ and if $\ell < n$, there is an infinite number of solutions.

Thus, the method not only decides whether or not a system has a solution, but if a solution exists, it gives a practical method for obtaining that solution. As a result of the above we can now state

COROLLARY 2   *A system of linear equations is consistent if and only if the reduced echelon matrix of its augmented matrix has no leading element in its last column.*

COROLLARY 3   *If $m < n$, then a homogeneous system of m linear equations in n variables has at least one non-trivial solution. If $\ell = n$, the trivial solution is the unique solution.*

COROLLARY 4   *A system of n linear homogeneous equations in n variables has a non-trivial solution if and only if its reduced echelon matrix $R \neq I_n$, where I_n is the $n \times n$ matrix with 1 in the (i, i)-positions $(i = 1, \ldots, n)$ with zeros elsewhere.*

PROOF   If $R = I_n$, then the system reduces to $x_1 = \ldots = x_n = 0$, i.e. we have the trivial solution.

If $R \neq I_n$, then R has a row and column of zeros, i.e. $\ell < n$ and by Corollary 3, the system has a non-trivial solution. ∎

11

EXAMPLES

1. Solve completely the following system of linear equations

$$5x_2 + 35x_3 - 24x_4 = 1$$
$$2x_1 + x_2 - x_3 + x_4 = 0$$
$$3x_1 + 2x_2 + 2x_3 - x_4 = 1$$
$$5x_1 + 3x_2 + x_3 = 1$$

The augmented matrix of this system is the matrix (2) of the last example in §1.2, which is row equivalent to the matrix (3). Thus, this system of linear equations is equivalent to the system

$$x_1 - 4x_3 = 26$$
$$x_2 + 7x_3 = -43$$
$$x_4 = -9$$

or

$$x_1 = 4x_3 + 26, x_2 = -7x_3 - 43, x_4 = -9$$

Put $x_3 = \lambda$, then the general solution is

$$(x_1, x_2, x_3, x_4) = (4\lambda + 26, -7\lambda - 43, \lambda, -9)$$

where λ may be given arbitrary values.

2. Solve completely the following system of linear equations

$$x_1 - x_2 + x_3 = 0$$
$$x_1 + x_2 + 2x_3 = 0$$
$$x_1 + 2x_2 - x_3 = 0$$

The matrix of coefficients of this system is

$$\begin{pmatrix} 1 & -1 & 1 \\ 1 & 1 & 2 \\ 1 & 2 & -1 \end{pmatrix}$$

By carrying out the following elementary row operations, we obtain its reduced echelon matrix

$$\begin{pmatrix} 1 & -1 & 1 \\ 1 & 1 & 2 \\ 1 & 2 & 1 \end{pmatrix} \xrightarrow[\substack{r_2 \to r_2 - r_1 \\ r_3 \to r_3 - r_1}]{} \begin{pmatrix} 1 & -1 & 1 \\ 0 & 2 & 1 \\ 0 & 3 & -2 \end{pmatrix}$$

$$\xrightarrow[\substack{r_2 \to \frac{1}{2}r_2}]{} \begin{pmatrix} 1 & -1 & 1 \\ 0 & 1 & \frac{1}{2} \\ 0 & 3 & -2 \end{pmatrix}, \xrightarrow[\substack{r_1 \to r_1 + r_2 \\ r_3 \to r_3 - 3r_2}]{}$$

$$\begin{pmatrix} 1 & 0 & \frac{3}{2} \\ 0 & 1 & \frac{1}{2} \\ 0 & 0 & -\frac{7}{2} \end{pmatrix} \xrightarrow[\substack{r_3 \to -\frac{2}{7}r_3}]{} \begin{pmatrix} 1 & 0 & \frac{3}{2} \\ 0 & 1 & \frac{1}{2} \\ 0 & 0 & 1 \end{pmatrix}$$

$$\xrightarrow[\substack{r_1 \to r_1 - \frac{3}{2}r_3 \\ r_2 \to r_2 - \frac{1}{2}r_3}]{} \begin{pmatrix} 1 & 0 & 0 \\ 0 & 1 & 0 \\ 0 & 0 & 1 \end{pmatrix}$$

which implies by Corollary 4 that the trivial solution $x_1 = x_2 = x_3 = 0$ is the unique solution.

3. For what value of k will the system

$$2x_1 + x_2 = 5$$
$$x_1 - 3x_2 = -1$$
$$3x_1 + 4x_2 = k$$

be consistent?

For that value of k, find the complete solution

$$\begin{pmatrix} 2 & 1 & | & 5 \\ 1 & -3 & | & -1 \\ 3 & 4 & | & k \end{pmatrix} \xrightarrow[\substack{r_1 \leftrightarrow r_2}]{} \begin{pmatrix} 1 & -3 & | & -1 \\ 2 & 1 & | & 5 \\ 3 & 4 & | & k \end{pmatrix}$$

$$\xrightarrow[\substack{r_2 \to r_2 - 2r_1 \\ r_3 \to r_3 - 3r_1}]{} \begin{pmatrix} 1 & -3 & | & -1 \\ 0 & 7 & | & 7 \\ 0 & 13 & | & k+3 \end{pmatrix} \xrightarrow[\substack{r_2 \to \frac{1}{7}r_2 \\ r_3 \to r_3 - 13r_2 \\ r_1 \to r_1 + 3r_2}]{}$$

$$\begin{pmatrix} 1 & 0 & | & 2 \\ 0 & 1 & | & 1 \\ 0 & 0 & | & k-10 \end{pmatrix}$$

13

Thus, by Corollary 2, the system of linear equations is consistent if and only if $k - 10 = 0$, i.e. $k = 10$. If $k = 10$, the reduced echelon matrix is

$$\begin{pmatrix} 1 & 0 & | & 2 \\ 0 & 1 & | & 1 \\ 0 & 0 & | & 0 \end{pmatrix}$$

and the solution is $x_1 = 2$, $x_2 = 1$.

Exercises 1.3

1. Solve the homogeneous systems of linear equations with the following matrices of coefficients

(i) $\begin{pmatrix} 2 & -1 & 1 & 0 & 1 \\ 1 & 2 & -1 & 2 & 1 \\ 1 & -8 & 5 & -6 & -1 \end{pmatrix}$

(ii) $\begin{pmatrix} 1 & -1 & 0 & 2 & 3 \\ 5 & 2 & 1 & 0 & 1 \\ -1 & 1 & 2 & -3 & 4 \end{pmatrix}$

(iii) $\begin{pmatrix} 1 & 0 & -1 & 1 \\ 2 & 1 & -1 & 1 \\ -1 & 2 & 0 & 1 \\ 1 & 0 & 2 & -3 \end{pmatrix}$

(iv) $\begin{pmatrix} 5 & -3 & 2 & 1 & 1 & 1 \\ 1 & 0 & -1 & 2 & 1 & -1 \\ 2 & -3 & 5 & -5 & -4 & 6 \\ 3 & -6 & 11 & -12 & -5 & 9 \\ 7 & -3 & 0 & 5 & 3 & -1 \end{pmatrix}$

2. Solve the non-homogeneous systems of linear equations with the following augmented matrices

(i) $\begin{pmatrix} 2 & -1 & 0 & 1 & | & -1 \\ -1 & 1 & 2 & 1 & | & 5 \\ 1 & 0 & 2 & 2 & | & 4 \end{pmatrix}$

(ii) $\begin{pmatrix} 3 & 1 & -1 & 1 & | & 2 \\ -1 & 0 & 2 & 5 & | & 3 \\ 2 & 1 & 1 & 6 & | & 2 \end{pmatrix}$

14

(iii) $\begin{pmatrix} 2 & 1 & -1 & 1 & 1 & | & 2 \\ 5 & 4 & 2 & 4 & 4 & | & 8 \\ 0 & 1 & 1 & 0 & 2 & | & 3 \\ -1 & 1 & 5 & 1 & 1 & | & 2 \end{pmatrix}$

(iv) $\begin{pmatrix} 2 & 0 & 1 & -1 & 1 & | & 4 \\ -1 & 0 & 2 & 3 & 4 & | & 1 \\ 1 & 1 & 0 & -1 & 1 & | & 2 \\ 1 & 1 & 5 & 4 & 10 & | & 7 \end{pmatrix}$

3. Find the conditions which λ and μ must satisfy for the following systems of linear equations to have (i) a unique solution, (ii) no solution, (iii) an infinite number of solutions

(a) $2x + 3y + z = 5$

$3x - y + \lambda z = 2$

$x + 7y - 6z = \mu$

(b) $x + y - 4z = 0$

$2x + 3y + z = 1$

$4x + 7y + \lambda z = \mu$

4. For what values of λ have the equations

$x + y + z = a$

$\lambda x + 2y + z = b$

$\lambda^2 x + 4y + z = c$

a unique solution? In the exceptional cases, find the conditions to be satisfied by a, b, c in order that a solution may exist and find the general solution.

5. Determine the values of λ for which the following systems of equations are consistent and for those values of λ find the complete solutions

(i) $5x + 2y - z = 1$

$2x + 3y + 4z = 7$

$4x - 5y + \lambda z = \lambda - 5$

(ii) $x + 2y + \lambda z = 0$

$2x + 3y - 2z = \lambda$

$\lambda x + y + \lambda^2 z = 3$

(iii) $\quad x + 5y + 3 = 0$

$\qquad 5x + y - \lambda = 0$

$\qquad x + 2y + \lambda = 0$

(iv) $2x + 3y + z + t = 0$

$\qquad x + 2y + z - t = 1$

$\qquad 3x + 5y + 2z + t = a$

$\qquad 6x + 10y + 4z + t = (\lambda + 1)^2$

6. Find the complete solution of the system of equations

$$y + z + u + 2v = 2$$

$$-x + 4y + 3z + 3u + 4v = 7$$

$$2x + y + 3z + 2u + 8v = 3$$

$$3x + y + 4z - u + 4v = 0$$

$$5x + 2y + 7z + 10v = 2$$

7. Solve completely the system of linear equations

$$x_1 + x_2 \qquad\qquad\qquad = 0$$

$$x_1 + x_2 + x_3 \qquad\qquad = 0$$

$$x_2 + x_3 + x_4 \qquad\qquad = 0$$

$$\cdots\cdots\cdots\cdots\cdots$$

$$x_{n-3} + x_{n-2} + x_{n-1} \qquad = 0$$

$$x_{n-2} + x_{n-1} + x_n = 0$$

$$x_{n-1} + x_n = 0$$

when (i) $n = 8$ (ii) $n = 9$.

1.4 Matrix Algebra

DEFINITION 1.8 *EQUALITY OF MATRICES* *If $A = (\alpha_{ij})$ is an $m \times n$ matrix and $B = (\beta_{ij})$ is a $p \times q$ matrix then $A = B$ if and only if $m = p$, $n = q$ and $\alpha_{ij} = \beta_{ij}$ ($i = 1, \ldots, m, j = 1, \ldots, n$).*
Zero Matrix *The $m \times n$ matrix $0 = (\alpha_{ij})$ such that $\alpha_{ij} = 0$ ($i = 1, \ldots, m; j = 1, \ldots, n$) is called the zero matrix.*
Matrix Addition *If $A = (\alpha_{ij})$ is an $m \times n$ matrix and $B = (\beta_{ij})$ is a $p \times q$ matrix then $A + B$ is defined if and only if $m = p$, $n = q$ and then*

$$A + B = (\alpha_{ij} + \beta_{ij})$$

is the $m \times n$ matrix obtained by adding the corresponding elements in A and B.

16

Scalar Multiplication   *If $A = (\alpha_{ij})$ is an $m \times n$ matrix and $\alpha \in K$, then*

$$\alpha A = (\alpha \alpha_{ij})$$

*is the $m \times n$ matrix obtained by multiplying each element of A by α; αA is called the **scalar multiple** of A by α.*

EXAMPLE

$$\begin{pmatrix} 1 & 2 & 5 \\ 3 & 4 & 6 \end{pmatrix} + \begin{pmatrix} 0 & 1 & -1 \\ 2 & 5 & -3 \end{pmatrix} = \begin{pmatrix} 1+0 & 2+1 & 5-1 \\ 3+2 & 4+5 & 6-3 \end{pmatrix}$$

$$= \begin{pmatrix} 1 & 3 & 4 \\ 5 & 9 & 3 \end{pmatrix}$$

but $\begin{pmatrix} 1 & 0 \\ 2 & 1 \end{pmatrix} + \begin{pmatrix} 1 & -1 & 1 \\ 2 & 3 & 5 \end{pmatrix}$ is not defined.

If $A = \begin{pmatrix} 1 & 2 & 5 \\ 3 & 4 & 6 \end{pmatrix}$ and $\alpha = -2$, then

$$-2A = \begin{pmatrix} -2 & -4 & -10 \\ -6 & -8 & -12 \end{pmatrix}$$

The definitions of matrix addition and scalar multiplication are the expected definitions for matrices, but as motivation for matrix multiplication, we consider the following.

Suppose that we have the following systems of linear equations,

$$
\begin{aligned}
y_1 &= \alpha_{11} x_1 + \alpha_{12} x_2 + \alpha_{13} x_3 \\
y_2 &= \alpha_{21} x_1 + \alpha_{22} x_2 + \alpha_{23} x_3
\end{aligned}
\tag{5}
$$

and

$$
\begin{aligned}
x_1 &= \beta_{11} z_1 + \beta_{12} z_2 \\
x_2 &= \beta_{21} z_1 + \beta_{22} z_2 \\
x_3 &= \beta_{31} z_1 + \beta_{32} z_2
\end{aligned}
\tag{6}
$$

Then, substituting (6) in (5) gives

$$
\begin{aligned}
y_1 &= \alpha_{11}(\beta_{11} z_1 + \beta_{12} z_2) + \alpha_{12}(\beta_{21} z_1 + \beta_{22} z_2) + \alpha_{13}(\beta_{31} z_1 + \beta_{32} z_2) \\
&= (\alpha_{11}\beta_{11} + \alpha_{12}\beta_{21} + \alpha_{13}\beta_{31}) z_1 + (\alpha_{11}\beta_{12} + \alpha_{12}\beta_{22} + \alpha_{13}\beta_{32}) z_2 \\
y_2 &= \alpha_{21}(\beta_{11} z_1 + \beta_{12} z_2) + \alpha_{23}(\beta_{21} z_1 + \beta_{22} z_2) + \alpha_{22}(\beta_{31} z_1 + \beta_{32} z_2) \\
&= (\alpha_{21}\beta_{11} + \alpha_{22}\beta_{21} + \alpha_{23}\beta_{31}) z_1 + (\alpha_{21}\beta_{12} + \alpha_{22}\beta_{22} + \alpha_{23}\beta_{32}) z_2
\end{aligned}
$$

17

or in other words, if $A = \begin{pmatrix} \alpha_{11} & \alpha_{12} & \alpha_{13} \\ \alpha_{21} & \alpha_{22} & \alpha_{23} \end{pmatrix}$ and $B = \begin{pmatrix} \beta_{11} & \beta_{12} \\ \beta_{21} & \beta_{22} \\ \beta_{31} & \beta_{32} \end{pmatrix}$

are the matrices of coefficients of the systems (5) and (6) respectively, then the matrix of coefficients of the resulting system, which we shall denote by AB, is

$$\begin{pmatrix} \alpha_{11}\beta_{11} + \alpha_{12}\beta_{21} + \alpha_{13}\beta_{31} & \alpha_{11}\beta_{12} + \alpha_{12}\beta_{22} + \alpha_{13}\beta_{32} \\ \alpha_{21}\beta_{11} + \alpha_{22}\beta_{21} + \alpha_{23}\beta_{31} & \alpha_{21}\beta_{12} + \alpha_{22}\beta_{22} + \alpha_{23}\beta_{32} \end{pmatrix}$$

That is, if A is a 2×3 matrix and B is a 3×2 matrix, then AB is a 2×2 matrix. Note, for example, that the $(1, 1)$-element in AB has been obtained by multiplying the elements of the first row of A with the corresponding elements in the first column of B and summing.

We use this now to give the following definition of matrix multiplication.

DEFINITION 1.9   *If A is an $m \times n$ matrix and B is a $p \times q$ matrix then AB is defined if and only if $n = p$ and then AB is the $m \times q$ matrix*

$$AB = \left(\sum_{k=1}^{n} \alpha_{ik}\beta_{kj} \right)$$

that is, the (i, j)-element of AB is obtained by multiplying the elements in the ith row of A with the corresponding elements in the jth column of B and summing.

i.e. if $A = \begin{pmatrix} \alpha_{11} & \alpha_{12} & \cdots & \alpha_{1n} \\ \alpha_{21} & \alpha_{22} & \cdots & \alpha_{2n} \\ \cdot & \cdot & & \cdot \\ \cdot & \cdot & & \cdot \\ \cdot & \cdot & & \cdot \\ \alpha_{m1} & \alpha_{m2} & \cdots & \alpha_{mn} \end{pmatrix}$ and $B = \begin{pmatrix} \beta_{11} & \beta_{12} & \cdots & \beta_{1q} \\ \beta_{21} & \beta_{22} & \cdots & \beta_{2q} \\ \cdot & \cdot & & \cdot \\ \cdot & \cdot & & \cdot \\ \cdot & \cdot & & \cdot \\ \beta_{n1} & \beta_{n2} & \cdots & \beta_{nq} \end{pmatrix}$

then

$$AB = \begin{pmatrix} \alpha_{11}\beta_{11} + \alpha_{12}\beta_{21} + \ldots + \alpha_{1n}\beta_{n1} & \alpha_{11}\beta_{12} + \alpha_{12}\beta_{22} + \ldots + \alpha_{1n}\beta_{n2} \cdot \\ \alpha_{21}\beta_{11} + \alpha_{22}\beta_{21} + \ldots + \alpha_{2n}\beta_{n1} & \alpha_{21}\beta_{12} + \alpha_{22}\beta_{22} + \ldots + \alpha_{2n}\beta_{n2} \cdot \end{pmatrix}$$

The system of m linear equations in n variables $\sum\limits_{j=1}^{n} \alpha_{ij}\, x_j = \beta_i$
$(i = 1, \ldots, m)$ can be expressed in matrix form as $AX = b$, where

$$A = (\alpha_{ij}),\ X = \begin{pmatrix} x_1 \\ \cdot \\ \cdot \\ \cdot \\ x_n \end{pmatrix},\ b = \begin{pmatrix} \beta_1 \\ \cdot \\ \cdot \\ \beta_m \end{pmatrix}$$

EXAMPLE

Let $A = \begin{pmatrix} 2 & 5 & 3 \\ 1 & 2 & -1 \end{pmatrix}$ and $B = \begin{pmatrix} 1 & 0 \\ -1 & 1 \\ 2 & 1 \end{pmatrix}$ then AB is defined and

is the 2×2 matrix

$$\begin{pmatrix} 2.1 + 5.-1 + \ 3.2 & 2.0 + 5.1 + \ 3.1 \\ 1.1 + 2.-1 + -1.2 & 1.0 + 2.1 + -1.1 \end{pmatrix} = \begin{pmatrix} 3 & 8 \\ -3 & 1 \end{pmatrix}$$

BA is also defined and is the 3×3 matrix

$$\begin{pmatrix} 1.2 + 0.1 & 1.5 + 0.2 & 1.3 + 0.-1 \\ -1.2 + 1.1 & -1.5 + 1.2 & -1.3 + 1.-1 \\ 2.2 + 1.1 & 2.5 + 1.2 & 2.3 + 1.-1 \end{pmatrix} = \begin{pmatrix} 2 & 5 & 3 \\ -1 & -3 & -4 \\ 5 & 12 & 5 \end{pmatrix}$$

We note that AB and BA are not even of the same shape and thus in general $AB \neq BA$. Even if A and B are of the same shape, $AB \neq BA$, in general, for example

if $A = \begin{pmatrix} 1 & 0 \\ -1 & 1 \end{pmatrix}$ and $B = \begin{pmatrix} 1 & 1 \\ 0 & 1 \end{pmatrix}$, then

$$AB = \begin{pmatrix} 1 & 1 \\ -1 & 0 \end{pmatrix} \text{ and } BA = \begin{pmatrix} 0 & 1 \\ -1 & 1 \end{pmatrix}$$

We have seen in the above example that one of the laws of algebra, namely the commutative rule of multiplication is not satisfied. In the remainder of this book, a great deal of our work will depend on the manipulation of matrices and it would be of interest whether or not there are any further restrictions on the normal manipulations allowed in algebra. The following theorem shows that there are not.

THEOREM 1.10   *If $A = (\alpha_{ij}), B = (\beta_{ij}), C = (\gamma_{ij})$ are $m \times n$, $s \times t$ and $p \times q$ matrices respectively, then*

(i) (*Commutative Law of Addition*)
     $A + B = B + A$, where both sides are defined if and only if $m = s$,
     $n = t$.

(ii) (*Associative Law of Addition*)
     $(A + B) + C = A + (B + C)$ where both sides are defined if and
     only if $m = s = p$, $n = t = q$.

(iii) $A + 0 = A$.

(iv) $A + (-A) = 0$.

(v) (*Associative Law of Multiplication*)
     $A(BC) = (AB)C$ where both sides are defined if and only if $n = s$,
     $t = p$.

(vi) (*Left and Right Distributive Laws*)
     $A(B + C) = AB + AC$ where both sides are defined if and only if
     $n = s = p$, $t = q$
     $(B + C)A = BA + CA$ where both sides are defined if and only if
     $s = p$, $t = q = m$.

PROOF   We shall only prove (v), the remaining parts are proved using
a similar method.

AB is defined if and only if $n = s$ and is an $m \times t$ matrix.

$(AB)C$ is defined if and only if $n = s$, $t = p$ and is an $m \times q$ matrix.

BC is defined if and only if $t = p$ and is an $s \times q$ matrix.

$A(BC)$ is defined if and only if $n = s$, $t = p$ and is an $m \times q$ matrix.

Thus we see that $(AB)C$ and $A(BC)$ are defined under the same
conditions and have the same shape. Using the rules for matrix
multiplication, we now further see that

$$AB = \left(\sum_{k=1}^{n} \alpha_{ik} \beta_{kj} \right)_{m \times t}$$

$$(AB)C = \left(\sum_{\ell=1}^{t} \left(\sum_{k=1}^{n} \alpha_{ik} \beta_{k\ell} \right) \gamma_{\ell j} \right)_{m \times q}$$

$$BC = \left(\sum_{\ell=1}^{t} \beta_{i\ell} \gamma_{\ell j} \right)_{s \times q}$$

$$A(BC) = \left(\sum_{k=1}^{n} \alpha_{ik} \left(\sum_{\ell=1}^{t} \beta_{k\ell} \gamma_{\ell j} \right) \right)_{m \times q}$$

$$= \left(\sum_{\ell=1}^{t} \sum_{k=1}^{n} (\alpha_{ik} \beta_{k\ell}) \gamma_{\ell j} \right)_{m \times q}$$

$$= (AB)C$$

20

But the reader should be further warned concerning manipulations of matrices, as illustrated by the following examples

(i) $\begin{pmatrix} 1 & 1 \\ 2 & 2 \end{pmatrix} \begin{pmatrix} 1 & 2 \\ -1 & -2 \end{pmatrix} = \begin{pmatrix} 0 & 0 \\ 0 & 0 \end{pmatrix}$

that is $AB = 0$, but neither A nor B is 0.

(ii) $AB = AC \nRightarrow B = C$.

i.e. $A(B - C) = 0$ can be true with $A \neq 0$ and $B \neq C$.

Exercises 1.4

1. Add and multiply the following matrices (if possible)

$$\begin{pmatrix} 1 & 2 & 3 \\ -1 & 1 & 2 \\ 2 & 1 & 3 \end{pmatrix} \begin{pmatrix} -1 & 0 & 1 \\ 5 & 1 & 2 \\ 1 & 1 & 2 \end{pmatrix} \begin{pmatrix} -1 & 2 & 4 \\ 2 & 0 & 3 \\ -1 & 1 & 3 \end{pmatrix} \begin{pmatrix} -1 & 2 & 3 \\ 3 & 2 & 1 \end{pmatrix}$$

$$\begin{pmatrix} -5 & 2 \\ 4 & 0 \\ 3 & 1 \end{pmatrix}$$

Also test the associative and distributive laws on these matrices.

2. If the matrix $A = \begin{pmatrix} a & b \\ c & d \end{pmatrix}$ commutes with the matrix $B = \begin{pmatrix} 1 & 0 \\ 0 & 0 \end{pmatrix}$, show that $b = c = 0$. Hence show that if A commutes with *every* 2×2 matrix, it has the form

$$A = \begin{pmatrix} a & 0 \\ 0 & a \end{pmatrix}$$

3. Use the fact that a square matrix X commutes with X^2 to show that if

$$X^2 = \begin{pmatrix} 4 & 1 & 0 \\ 0 & 4 & 1 \\ 0 & 0 & 4 \end{pmatrix}$$

then X is of the form $\begin{pmatrix} a & b & c \\ 0 & a & b \\ 0 & 0 & a \end{pmatrix}$. Hence, find all the matrices X which satisfy the above equation.

4. If $A = \begin{pmatrix} 7 & 4 \\ -9 & -5 \end{pmatrix}$, prove that for positive integral n,

$$A^n = \begin{pmatrix} 1 + 6n & 4n \\ -9n & 1 - 6n \end{pmatrix}$$

Verify that the result is true when n is a negative integer.

5. Find the most general real $n \times n$ matrix which commutes with the matrix

$$\begin{pmatrix} 0 & 0 & \ldots & 0 & 1 \\ 0 & 0 & \ldots & 1 & 0 \\ \cdot & & & & \\ \cdot & & & & \\ \cdot & & & & \\ 0 & 1 & \ldots & 0 & 0 \\ 1 & 0 & \ldots & 0 & 0 \end{pmatrix}$$

1.5 Special Types of Matrices

1. IDENTITY MATRIX

The $n \times n$ matrix $\begin{pmatrix} 1 & 0 & 0 & \ldots & 0 \\ 0 & 1 & 0 & \ldots & 0 \\ \cdot & & & & \cdot \\ \cdot & & & 1 & \cdot \\ 0 & 0 & \cdot & & 0 & 1 \end{pmatrix}$ is called the **identity matrix**,

denoted by I_n or I, that is $I = (\alpha_{ij})$, where $\alpha_{ii} = 1$, $\alpha_{ij} = 0$ $(i \neq j)$. It has the property that

$$AI_n = I_n A = A$$

for all $n \times n$ matrices A.

2. DIAGONAL MATRIX

The $n \times n$ matrix $\begin{pmatrix} \lambda_1 & 0 & \ldots & 0 \\ 0 & \lambda_2 & & \cdot \\ \cdot & & & \cdot \\ \cdot & & & \cdot \\ \cdot & & & \cdot \\ 0 & 0 & \ldots & \lambda_n \end{pmatrix}$ is called a **diagonal matrix** and

and denoted by diag $(\lambda_1, \ldots, \lambda_n)$.

3. INVERSE MATRIX

If $A = (\alpha_{ij})$ is an $n \times n$ matrix then an $n \times n$ matrix B such that

$$AB = BA = I_n$$

*is called an **inverse** of the matrix A.*

LEMMA 1.11   *The inverse of a matrix is unique.*

PROOF   Let C also be an inverse of A, then

$$AC = CA = I_n$$

and

$$C = CI_n = C(AB) = (CA)B = I_n B = B \qquad \blacksquare$$

The unique inverse of A is denoted by A^{-1}.

A matrix A which has an inverse is called a **non-singular** or **invertible** matrix, and if it has no inverse, it is called a **singular** or **non-invertible** matrix.

LEMMA 1.12   *If A and B are invertible $n \times n$ matrices, then AB is invertible and $(AB)^{-1} = B^{-1} A^{-1}$.*

PROOF   If A and B are invertible, then $AA^{-1} = I_n = A^{-1}A$ and $BB^{-1} = I_n = B^{-1}B$. Now

$$(AB)(B^{-1}A^{-1}) = I_n = (B^{-1}A^{-1})(AB)$$

and so AB is invertible and since this inverse is unique by Lemma 1.11, we have

$$(AB)^{-1} = B^{-1}A^{-1} \qquad \blacksquare$$

Later in this section a method of calculating the inverse of a matrix will be given and for determining when a matrix is non-singular.

4. TRANSPOSE OF A MATRIX

If $A = (\alpha_{ij})$ is an $m \times n$ matrix, the **transpose** of A is the $n \times m$ matrix denoted by A^t obtained by interchanging the rows and columns of A, that is

$$A^t = (\beta_{ij})$$

where $\beta_{ij} = \alpha_{ji}$ $(i = 1, \ldots, n; j = 1, \ldots, m)$.

EXAMPLE

If $A = \begin{pmatrix} 1 & 0 & -3 \\ 2 & 1 & 3 \end{pmatrix}$ then $A^t = \begin{pmatrix} 1 & 2 \\ 0 & 1 \\ -3 & 3 \end{pmatrix}$

The following will be useful later.

LEMMA 1.13 *If A and B are m × n and p × q matrices respectively, then*

(i) $(A + B)^t = A^t + B^t$, *where both sides are defined if and only if* $m = p$, $n = q$.

(ii) $(AB)^t = B^t A^t$, *where both sides are defined if and only if* $n = p$.

PROOF We shall prove (ii) only.

AB is defined if and only if $n = p$ and it is an $m \times q$ matrix. Thus, $(AB)^t$ is defined if and only if $n = p$ and $(AB)^t$ is a $q \times m$ matrix. Similarly $B^t A^t$ is defined under the same conditions.

If $A = (\alpha_{ij})_{m \times n}$, $B = (\beta_{ij})_{n \times q}$ then

$$(AB)^t = \left(\sum_{k=1}^{n} \alpha_{ik} \beta_{kj} \right)^t$$

$$= (\gamma_{ij})_{q \times m}$$

where $\gamma_{ij} = \sum_{k=1}^{n} \alpha_{jk} \beta_{ki}$.

If $A^t = (\alpha_{ij}')_{n \times m}$, $B^t = (\beta_{ij}')_{q \times n}$, then

$$B^t A^t = \left(\sum_{k=1}^{n} \beta_{ik}' \alpha_{kj}' \right)_{q \times m}$$

$$= \left(\sum_{k=1}^{n} \alpha_{jk} \beta_{ki} \right)_{q \times m}$$

$$= (AB)^t, \text{ as required} \qquad \blacksquare$$

COROLLARY *If* $A_1, A_2, \ldots, A_k$ *are k matrices such that* A_{i+1} *has the same number of rows as* A_i *has columns* $(i = 1, \ldots, k-1)$, *then* $(A_1 A_2 \ldots A_k)^t = A_k^t A_{k-1}^t \ldots A_1^t$.

PROOF Use induction on k. $\qquad \blacksquare$

LEMMA 1.14 *If A is an* $n \times n$ *matrix then* A^t *is invertible if and only if A is invertible. Further* $(A^t)^{-1} = (A^{-1})^t$.

PROOF If A is invertible then $AA^{-1} = I_n = A^{-1}A$, transposing gives $(A^{-1})^t A^t = I_n = A^t(A^{-1})^t$ and so A^t is invertible. The proof of the converse is similar. The above also implies by Lemma 1.11 that $(A^t)^{-1} = (A^{-1})^t$. ∎

5. SYMMETRIC, SKEW-SYMMETRIC AND ORTHOGONAL MATRICES

An $n \times n$ matrix A is called
  (i) a *symmetric matrix* if $A^t = A$
  (ii) a *skew-symmetric matrix* if $A^t = -A$
  (iii) an *orthogonal matrix* if $A^tA = AA^t = I_n$.

EXAMPLES

$$\begin{pmatrix} a & b & c \\ b & d & e \\ c & e & f \end{pmatrix} \text{ is a symmetric matrix}$$

and $\begin{pmatrix} 0 & a & b \\ -a & 0 & c \\ -b & -c & 0 \end{pmatrix}$ is a skew-symmetric matrix

and $\frac{1}{3}\begin{pmatrix} 1 & 2 & 2 \\ 2 & 1 & -2 \\ 2 & -2 & 1 \end{pmatrix}$ is an orthogonal matrix

Remarks

1. In a symmetric matrix, the elements are symmetric about the diagonal of the matrix.
2. In a skew-symmetric matrix, the diagonal elements are all zero.
3. If A is an orthogonal matrix, then A^t is its inverse.

Exercises 1.5

1. If A is an $n \times n$ matrix, prove that
   (i) A^tA is a symmetric matrix,
   (ii) $A + A^t$ is a symmetric matrix and $A - A^t$ is a skew-symmetric matrix
   (iii) A is the sum of a symmetric and a skew-symmetric matrix.

2. If A and B are symmetric and skew-symmetric matrices respectively and $AB = BA$ and $A + B$ is non-singular, prove that the matrix $(A + B)^{-1}(A - B)$ is orthogonal.

3. Show that, if $I + SA$ is non-singular, where A is a symmetric matrix and S is a skew-symmetric matrix, then the matrix

$$L = (I - SA)(I + SA)^{-1}$$

is such that $L^t AL = A$. Conversely, if $L^t AL = A$, where A is symmetric and $I + L$ and A are non-singular show that $S = (I + L)^{-1}(I - L)A^{-1}$ is skew-symmetric.

4. If
$$A(\alpha) = \begin{pmatrix} 1 & \alpha & \alpha^2/2 \\ 0 & 1 & \alpha \\ 0 & 0 & 1 \end{pmatrix}$$

show that $A(\alpha)A(\beta) = A(\alpha + \beta)$. Hence, find the inverse of $A(\alpha)$. Show that

$$A(3\alpha) - 3A(2\alpha) + 3A(\alpha) = I$$

and hence find a cubic equation satisfied by $A(\alpha)$.

5. If A is a real matrix, express the sum of the diagonal elements of $A^t A$ in terms of elements of A. Hence, show that if A, B are real and symmetric and C is real and skew-symmetric then $A^2 + B^2 = C^2$ implies $A = B = C = 0$. Does this conclusion still hold if A is not necessarily symmetric?

6. Prove that every 2×2 matrix X such that $X^t AX = B$, where
$A = \begin{pmatrix} 1 & 0 \\ 0 & -1 \end{pmatrix}$, $B = \begin{pmatrix} 0 & 1 \\ 1 & 0 \end{pmatrix}$ has one of the forms

$$\begin{pmatrix} \alpha & \tfrac{1}{2}\alpha^{-1} \\ \alpha & -\tfrac{1}{2}\alpha^{-1} \end{pmatrix} \quad \text{or} \quad \begin{pmatrix} \alpha & \tfrac{1}{2}\alpha^{-1} \\ -\alpha & \tfrac{1}{2}\alpha^{-1} \end{pmatrix}$$

Find all the matrices X which satisfy the additional relation $X^t X = I$

1.6 Elementary Matrices

DEFINITION 1.15 *An $n \times n$ matrix is called an **elementary matrix** if it can be obtained by applying an elementary row operation to I_n.*

Thus, elementary matrices are one of the following three types corresponding to the elementary row operations $r_i \to \alpha r_i, r_i \to r_j$, $r_i \to r_i + \alpha r_j$ respectively:

$$M_i(\alpha) = \begin{array}{c} \\ \\ \\ \\ i \\ \\ \\ \\ \end{array} \overset{\displaystyle i}{\begin{pmatrix} 1 & & & & & & \\ & \ddots & & & & & \\ & & 1 & & & & \\ & & & \alpha & & & \\ & & & & 1 & & \\ & & & & & \ddots & \\ & & & & & & 1 \end{pmatrix}} \qquad (i = 1, \ldots, n)$$

$$H_{ij} = \begin{array}{c} \\ \\ \\ i \\ \\ \\ j \\ \\ \\ \end{array} \begin{pmatrix} 1 & & & & & & & \\ & \ddots & & & & & & \\ & & 1 & & & & & \\ & & & 0 & 1 & & & \\ & & & 1 & & & & \\ & & & & & \ddots & & \\ & & & & & & 1 & \\ & & 1 & & & & 0 & \\ & & & & & & & 1 \\ & & & & & & & & \ddots \\ & & & & & & & & & 1 \end{pmatrix} \qquad (i,j = 1, \ldots, n)$$

$$A_{ij}(\alpha) = \begin{array}{c} \\ \\ i \\ \\ j \\ \\ \\ \end{array} \begin{pmatrix} 1 & & & & & & \\ & \ddots & & & & & \\ & & 1 & & \alpha & & \\ & & & \ddots & & & \\ & & & & 1 & & \\ & & & & & \ddots & \\ & & & & & & 1 \end{pmatrix} \qquad (i,j = 1, \ldots, n)$$

where every other non-diagonal element is zero.

LEMMA 1.16   (i) *If an m × n matrix B can be obtained from an m × n matrix A by applying an elementary row operation, then B is equal to the product of the corresponding m × m elementary matrix with A i.e. if e is the elementary row operation*

$$B = e(A) = e(I_m)A$$

(ii) *Every elementary matrix is invertible, the inverse of each elementary matrix is elementary.*

PROOF   (i) and (ii) are proved by considering each separate type.

(i) $r_i \to \alpha r_i$ then

$$M_i(\alpha)A = \begin{pmatrix} \alpha_{11} & \cdots & \alpha_{1n} \\ \vdots & & \vdots \\ \alpha\alpha_{i1} & \cdots & \alpha\alpha_{in} \\ \vdots & & \vdots \\ \alpha_{m1} & \cdots & \alpha_{mn} \end{pmatrix}$$

or

$$\begin{pmatrix} 1 & & & & & \\ & 1 & & & & \\ & & \ddots & & & \\ & & & \alpha & & \\ & & & & \ddots & \\ & & & & & 1 \end{pmatrix} \begin{pmatrix} \alpha_{11} & \cdots & \alpha_{1n} \\ \vdots & & \vdots \\ \alpha_{i1} & \cdots & \alpha_{in} \\ \vdots & & \vdots \\ \alpha_{m1} & \cdots & \alpha_{mn} \end{pmatrix} = \begin{pmatrix} \alpha_{11} & \cdots & \alpha_{1n} \\ \vdots & & \vdots \\ \alpha\alpha_{i1} & & \alpha\alpha_{in} \\ \vdots & & \vdots \\ \alpha_{m1} & \cdots & \alpha_{mn} \end{pmatrix}$$

and similarly

$$H_{ij}A = \begin{pmatrix} \alpha_{11} & \cdots & \alpha_{1n} \\ \vdots & & \vdots \\ \alpha_{j1} & \cdots & \alpha_{jn} \\ \vdots & & \vdots \\ \alpha_{i1} & \cdots & \alpha_{in} \\ \vdots & & \vdots \\ \alpha_{m1} & \cdots & \alpha_{mn} \end{pmatrix}$$

and $$A_{ij}(\alpha)A = \begin{pmatrix} \alpha_{11} & \cdots & \alpha_{1n} \\ \vdots & & \vdots \\ \alpha_{i1} + \alpha\alpha_{j1} & \cdots & \alpha_{in} + \alpha\alpha_{jn} \\ \vdots & & \vdots \\ \alpha_{m1} & \cdots & \alpha_{mn} \end{pmatrix}$$

28

(ii) For this part verify that

$$(M_i(\alpha))^{-1} = M_i\left(\frac{1}{\alpha}\right), \quad (H_{ij})^{-1} = H_{ij}$$

and $(A_{ij}(\alpha))^{-1} = A_{ij}(-\alpha)$ ∎

EXAMPLE

If $\quad A = \begin{pmatrix} 1 & 0 & 1 \\ 2 & -1 & 0 \\ 1 & 1 & 0 \end{pmatrix} \xrightarrow{\;r_2 \to r_2 - 2r_1\;} \begin{pmatrix} 1 & 0 & 1 \\ 0 & -1 & -2 \\ 1 & 1 & 0 \end{pmatrix}$

then $\quad \begin{pmatrix} 1 & 0 & 1 \\ 0 & -1 & -2 \\ 1 & 1 & 0 \end{pmatrix} = \begin{pmatrix} 1 & 0 & 0 \\ -2 & 1 & 0 \\ 0 & 0 & 1 \end{pmatrix} \begin{pmatrix} 1 & 0 & 1 \\ 2 & -1 & 0 \\ 1 & 1 & 0 \end{pmatrix}$

Also $\quad \begin{pmatrix} 1 & 0 & 0 \\ -2 & 1 & 0 \\ 0 & 0 & 1 \end{pmatrix} \begin{pmatrix} 1 & 0 & 0 \\ 2 & 1 & 0 \\ 0 & 0 & 1 \end{pmatrix} = \begin{pmatrix} 1 & 0 & 0 \\ 0 & 1 & 0 \\ 0 & 0 & 1 \end{pmatrix}$

We are now ready to prove the main result in this section.

THEOREM 1.17 *If an m × n matrix A is row equivalent to an m × n matrix B, then there exist a finite number of elementary matrices $E_1, E_2, \ldots, E_k$ such that*

$$B = E_1 E_2 \ldots E_k A$$

PROOF   In Theorem 1.5, we saw that B can be obtained from A by a finite, say k, sequence of elementary row operations. The proof is by induction on k. If $k = 1$, then by Lemma 1.16 (i) $B = E_1 A$ for some elementary matrix E_1.
If $k > 1$, we assume that if a matrix B' can be obtained from A by $k - 1$ elementary row operations, then there exist elementary matrices $E_2, \ldots, E_k$ such that

$$B' = E_2 \ldots E_k A$$

Now B is obtained from B' by applying one further elementary row operation and by Lemma 1.16 (i)

$$B = E_1 B'$$

for some elementary row operation E_1 and thus

$$B = E_1(E_2 \ldots E_k A) = E_1 E_2 \ldots E_k A$$

as required. ∎

There are two important corollaries to this theorem.

COROLLARY 1   *If R is the reduced echelon matrix of A, then there exist a finite number of elementary matrices $E_1, \ldots, E_k$ such that*

$$A = E_1 E_2 \ldots E_k R$$

PROOF   By Theorem 1.17 there exist elementary matrices $E_1', \ldots, E_k'$ such that

$$R = E_1' E_2' \ldots E_k' A$$

But, by lemma 1.16 (ii), each elementary matrix is non-singular and its inverse is also an elementary matrix. Thus

$$A = E_k'^{-1} E_{k-1}'^{-1} \ldots E_1'^{-1} R$$

as required. ∎

COROLLARY 2   *The following statements are equivalent.*
  (i) *the $n \times n$ matrix A is invertible.*
  (ii) *the homogeneous system $AX = 0$ of n linear equations in n variables $x_1, \ldots, x_n$ has no non-trivial solution.*
  (iii) *A is a product of elementary matrices.*

PROOF   (i) $\Rightarrow$ (ii) If A is invertible, the system of n linear equations in n variables given by $AX = 0$ has no non-trivial solution, since then

$$X = (A^{-1}A)X = A^{-1}(AX) = A^{-1}(0) = 0$$

(ii) $\Rightarrow$ (iii) By Corollary 4 to Theorem 1.7 the reduced echelon matrix of A is I_n and by Corollary 1 above

$$A = E_1 \ldots E_k$$

is a product of elementary matrices.

  (iii) $\Rightarrow$ (i) If $A = E_1 E_2 \ldots E_k$, where each E_i is an elementary matrix, then since elementary matrices are invertible, then by Lemma 1.12

$$A^{-1} = E_k^{-1} \ldots E_1^{-1}$$

and A is invertible. ∎

Hence in order to determine whether a matrix is invertible or not we note that if the reduced echelon matrix of A is R and

30

(i) if $R = I_n$, the matrix A is invertible.

(ii) if R contains a row of zeros, then A is not invertible.

This corollary is important in that it gives an explicit method for calculating the inverse of an invertible matrix. If A is non-singular, then the reduced echelon matrix of A is the identity matrix. If $E_1, \ldots, E_k$ are the elementary matrices which correspond to the elementary row operations which must be performed on A to give I_n

i.e. $\quad I_n = E_k \ldots E_2 E_1 A$

then $\quad A = E_1^{-1} E_2^{-1} \ldots E_k^{-1} I_n$

and $\quad A^{-1} = E_k \ldots E_2 E_1 I_n$

That is, A^{-1} is determined by applying the same elementary row operations to I_n as were used to obtain I_n from A.

EXAMPLES

1. Find the inverse of $\begin{pmatrix} 1 & -1 & 2 \\ 3 & 2 & 4 \\ 0 & 1 & -2 \end{pmatrix}$.

$$\left(\begin{array}{ccc|ccc} 1 & -1 & 2 & 1 & 0 & 0 \\ 3 & 2 & 4 & 0 & 1 & 0 \\ 0 & 1 & -2 & 0 & 0 & 1 \end{array}\right) \xrightarrow{r_2 \to r_2 - 3r_1} \left(\begin{array}{ccc|ccc} 1 & -1 & 2 & 1 & 0 & 0 \\ 0 & 5 & -2 & -3 & 1 & 0 \\ 0 & 1 & -2 & 0 & 0 & 1 \end{array}\right)$$

$$\xrightarrow{r_3 \leftrightarrow r_2} \left(\begin{array}{ccc|ccc} 1 & -1 & 2 & 1 & 0 & 0 \\ 0 & 1 & -2 & 0 & 0 & 1 \\ 0 & 5 & -2 & -3 & 1 & 0 \end{array}\right) \xrightarrow[r_3 \to r_3 - 5r_2]{r_1 \to r_1 + r_2}$$

$$\left(\begin{array}{ccc|ccc} 1 & 0 & 0 & 1 & 0 & 1 \\ 0 & 1 & -2 & 0 & 0 & 1 \\ 0 & 0 & 8 & -3 & 1 & -5 \end{array}\right) \xrightarrow{r_3 \to \frac{1}{8}r_3} \left(\begin{array}{ccc|ccc} 1 & 0 & 0 & 1 & 0 & 1 \\ 0 & 1 & -2 & 0 & 0 & 1 \\ 0 & 0 & 1 & -\frac{3}{8} & \frac{1}{8} & -\frac{5}{8} \end{array}\right)$$

$$\xrightarrow{r_2 \to r_2 + 2r_3} \left(\begin{array}{ccc|ccc} 1 & 0 & 0 & 1 & 0 & 1 \\ 0 & 1 & 0 & -\frac{6}{8} & \frac{2}{8} & -\frac{2}{8} \\ 0 & 0 & 1 & -\frac{3}{8} & \frac{1}{8} & -\frac{5}{8} \end{array}\right)$$

therefore $A^{-1} = \begin{pmatrix} 1 & 0 & 1 \\ -\frac{3}{4} & \frac{1}{4} & -\frac{1}{4} \\ -\frac{3}{8} & \frac{1}{8} & -\frac{5}{8} \end{pmatrix}$

2. Are the matrices

(i) $A = \begin{pmatrix} 1 & 0 & 1 \\ -1 & 1 & 2 \\ 1 & 1 & 4 \end{pmatrix}$ 　　(ii) $\begin{pmatrix} 1 & -1 & 1 \\ 1 & 2 & 1 \\ 0 & 1 & 1 \end{pmatrix}$

invertible?

(i) $\begin{pmatrix} 1 & 0 & 1 \\ -1 & 1 & 2 \\ 1 & 1 & 4 \end{pmatrix} \xrightarrow[r_3 \to r_3 - r_1]{r_2 \to r_2 + r_1} \begin{pmatrix} 1 & 0 & 1 \\ 0 & 1 & 3 \\ 0 & 1 & 3 \end{pmatrix}$

$\xrightarrow{r_3 \to r_3 - r_2} \begin{pmatrix} 1 & 0 & 1 \\ 0 & 1 & 3 \\ 0 & 0 & 0 \end{pmatrix}$

and thus A is not invertible.

(ii) $\begin{pmatrix} 1 & -1 & 1 \\ 1 & 2 & 1 \\ 0 & 1 & 1 \end{pmatrix} \xrightarrow{r_2 \to r_2 - r_1} \begin{pmatrix} 1 & -1 & 1 \\ 0 & 3 & 0 \\ 0 & 1 & 1 \end{pmatrix}$

$\xrightarrow[\substack{r_1 \to r_1 + r_2 \\ r_3 \to r_3 - r_2}]{r_2 \to \frac{1}{3} r_2} \begin{pmatrix} 1 & 0 & 1 \\ 0 & 1 & 0 \\ 0 & 0 & 1 \end{pmatrix} \xrightarrow{r_1 \to r_1 - r_3} \begin{pmatrix} 1 & 0 & 0 \\ 0 & 1 & 0 \\ 0 & 0 & 1 \end{pmatrix}$

and thus A is invertible.

Exercises 1.6

1. Find inverses (if they exist) of the following matrices

(i) $\begin{pmatrix} 1 & 0 & 1 \\ 2 & -1 & 1 \\ 1 & 2 & 1 \end{pmatrix}$ 　(ii) $\begin{pmatrix} 1 & 2 & 3 \\ 1 & 3 & 5 \\ 1 & 5 & 12 \end{pmatrix}$ 　(iii) $\begin{pmatrix} 1 & 3 & -1 \\ 3 & -2 & 7 \\ 4 & -1 & 1 \end{pmatrix}$

32

(iv) $\begin{pmatrix} 3 & -1 & 4 \\ 5 & 1 & -3 \\ 4 & -1 & 1 \end{pmatrix}$ (v) $\begin{pmatrix} 0 & -1 & 2 & 1 \\ -4 & 3 & -3 & 5 \\ 1 & 0 & 0 & -1 \\ -1 & 1 & 0 & 1 \end{pmatrix}$

(vi) $\begin{pmatrix} 1 & 2 & -3 & 1 \\ 2 & 4 & 1 & 3 \\ -1 & 1 & 1 & 0 \\ -2 & 2 & -5 & -1 \end{pmatrix}$ (vii) $\begin{pmatrix} 1 & i & 0 \\ i & -1 & 1+i \\ 1-i & 0 & 2 \end{pmatrix}$,

where $i = \sqrt{-1}$.

1.7 Elementary Column Operations and Equivalent Matrices

Most of the material in this chapter has been developed in terms of
elementary row operations. It is clear that the definition of elementary
row operations (Definition 1.2) and row equivalent matrices (Definition
1.3) may be adapted in an obvious way to define **elementary column
operations** on a matrix and **column equivalent matrices**.

In Theorem 1.17, we saw that if R is the reduced echelon matrix of
an $m \times n$ matrix A, then there exists an invertible matrix P such that

$$PA = R$$

Indeed, it is easy to calculate P, since by Theorem 1.17,

$$R = E_1 E_2 \ldots E_k A$$

where $E_k, E_{k-1}, \ldots, E_1$ are the elementary matrices which correspond
to the sequence of elementary row operations which are carried out
successively in order to obtain R from A. Thus, we have

$$P = E_1 E_2 \ldots E_k$$
$$= E_1(E_2(\ldots E_{k-1}(E_k I_m)) \ldots)$$

and the **matrix P is obtained by applying to I_m the same elementary row
operations as were used to obtain R from A** (c.f. the method given for
determining the inverse of an invertible matrix in §1.6).

33

EXAMPLE

$$\begin{pmatrix} 2 & -1 & 0 & 1 & | & 1 & 0 & 0 \\ 1 & 2 & 1 & -1 & | & 0 & 1 & 0 \\ 2 & 9 & 4 & -5 & | & 0 & 0 & 1 \end{pmatrix} \xrightarrow[\; r_3 \to r_3 - 2r_2 \;]{\; r_1 \to r_1 - 2r_2 \;}$$

$$\begin{pmatrix} 0 & -5 & -2 & 3 & | & 1 & -2 & 0 \\ 1 & 2 & 1 & -1 & | & 0 & 1 & 0 \\ 0 & 5 & 2 & -3 & | & 0 & -2 & 1 \end{pmatrix} \xrightarrow[\; r_1 \to -\frac{1}{5} r_1 \;]{\; r_3 \to r_3 - r_1 \;}$$

$$\begin{pmatrix} 0 & 1 & \frac{2}{5} & -\frac{3}{5} & | & -\frac{1}{5} & \frac{2}{5} & 0 \\ 1 & 2 & 1 & -1 & | & 0 & 1 & 0 \\ 0 & 0 & 0 & 0 & | & 1 & -4 & 1 \end{pmatrix} \xrightarrow[\; r_1 \to r_1 - 2r_2 \;]{\; r_1 \leftrightarrow r_2 \;}$$

$$\begin{pmatrix} 1 & 0 & \frac{1}{5} & \frac{1}{5} & | & \frac{2}{5} & \frac{1}{5} & 0 \\ 0 & 1 & \frac{2}{5} & -\frac{3}{5} & | & -\frac{1}{5} & \frac{2}{5} & 0 \\ 0 & 0 & 0 & 0 & | & 1 & -4 & 1 \end{pmatrix}$$

It is easily verified that

$$\begin{pmatrix} \frac{2}{5} & \frac{1}{5} & 0 \\ -\frac{1}{5} & \frac{2}{5} & 0 \\ 1 & -4 & 1 \end{pmatrix} \begin{pmatrix} 2 & -1 & 0 & 1 \\ 1 & 2 & 1 & -1 \\ 2 & 9 & 4 & -5 \end{pmatrix} = \begin{pmatrix} 1 & 0 & \frac{1}{5} & \frac{1}{5} \\ 0 & 1 & \frac{2}{5} & -\frac{3}{5} \\ 0 & 0 & 0 & 0 \end{pmatrix}$$

Theorem 1.5 and Theorem 1.17 have their equivalent statements in terms of elementary column operations. In particular, we can prove

THEOREM 1.18 *If an m × n matrix A is column equivalent to an m × n matrix B, then there exist a finite number of elementary matrices $E_1, E_2, \ldots, E_\varrho$ such that*

$$B = AE_1E_2 \ldots E_\varrho$$

Following the above argument, we see that there exists an invertible $n \times n$ matrix Q such that $B = AQ$.

We apply this, in particular, to the reduced echelon matrix R of an $m \times n$ matrix A. Suppose that the leading 1's appear in columns $j_1, j_2, \ldots, j_r$, then by subtracting suitable multiples of these columns from succeeding columns and then the column interchanges $c_1 \leftrightarrow c_{j_1}$, $c_2 \leftrightarrow c_{j_2}, \ldots, c_r \leftrightarrow c_{j_r}$, we have that R is column equivalent to the matrix

34

$$N = \left(\begin{array}{c|c} I_{r \times r} & 0_{r \times (n-r)} \\ \hline 0_{(m-r) \times r} & 0_{(m-r) \times (n-r)} \end{array} \right)$$

where $I_{r \times r}$ is the $r \times r$ identity matrix and $0_{p \times q}$ is the zero $p \times q$ matrix. In other words, we can state this as

THEOREM 1.19   *If A is an $m \times n$ matrix then there exists an invertible $m \times m$ matrix P and an invertible $n \times n$ matrix Q such that*

$$PAQ = N$$

where N is the matrix given above.

As explained above, the matrices P and Q are easily calculated as illustrated in the following:

EXAMPLE

If $A = \begin{pmatrix} 2 & -1 & 0 & 1 \\ 1 & 2 & 1 & -1 \\ 2 & 9 & 4 & -5 \end{pmatrix}$, we have seen in the preceding example

that if $P = \begin{pmatrix} \frac{2}{5} & \frac{1}{5} & 0 \\ -\frac{1}{5} & \frac{2}{5} & 0 \\ 1 & -4 & 1 \end{pmatrix}$, then

$$PA = \begin{pmatrix} 1 & 0 & \frac{1}{5} & \frac{1}{5} \\ 0 & 1 & \frac{2}{5} & -\frac{3}{5} \\ 0 & 0 & 0 & 0 \end{pmatrix}$$

Now, we have

$$\left(\begin{array}{cccc|cccc} 1 & 0 & \frac{1}{5} & \frac{1}{5} & 1 & 0 & 0 & 0 \\ 0 & 1 & \frac{2}{5} & -\frac{3}{5} & 0 & 1 & 0 & 0 \\ 0 & 0 & 0 & 0 & 0 & 0 & 1 & 0 \\ & & & & 0 & 0 & 0 & 1 \end{array} \right) \xrightarrow[\substack{c_3 \to c_3 - \frac{2}{5}c_2 \\ c_4 \to c_4 + \frac{3}{5}c_2}]{\substack{c_3 \to c_3 - \frac{1}{5}c_1 \\ c_4 \to c_4 - \frac{1}{5}c_1}}$$

$$\left(\begin{array}{cccc|cccc} 1 & 0 & 0 & 0 & 1 & 0 & -\frac{1}{5} & -\frac{1}{5} \\ 0 & 1 & 0 & 0 & 0 & 1 & -\frac{2}{5} & \frac{3}{5} \\ 0 & 0 & 0 & 0 & 0 & 0 & 1 & 0 \\ & & & & 0 & 0 & 0 & 1 \end{array} \right)$$

and if $Q = \begin{pmatrix} 1 & 0 & -\frac{1}{5} & -\frac{1}{5} \\ 0 & 1 & -\frac{2}{5} & \frac{3}{5} \\ 0 & 0 & 1 & 0 \\ 0 & 0 & 0 & 1 \end{pmatrix}$, then

$$PAQ = \begin{pmatrix} 1 & 0 & 0 & 0 \\ 0 & 1 & 0 & 0 \\ 0 & 0 & 0 & 0 \end{pmatrix}$$

This leads us to make the following

DEFINITION 1.20 *Two $m \times n$ matrices A and B are **equivalent** if there exist invertible matrices P and Q such that*

$$PAQ = B$$

This is easily shown to be an equivalence relation on the set of $m \times n$ matrices over a field K, since any $m \times n$ matrix is clearly equivalent to itself; if $PAQ = B$ then $P^{-1}BQ^{-1} = A$; and if $PAQ = B$ and $P'BQ' = C$ then $(P'P)A(QQ') = C$. What we have proved above is that every $m \times n$ matrix A is equivalent to a matrix with the simple form N, where r denotes the number of non-zero rows in the reduced echelon form of A. N may be regarded as a **canonical form** of A under this equivalence relation, that is, a representative of the equivalence class which contains A under this equivalence relation. Later, in Chapter IV (see Corollary to Theorem 4.20), it will be seen that r is the rank of the matrix A. N is called the *normal* form of the matrix A.

Exercise 1.7

1. Find invertible matrices P and Q such that PAQ is in normal form for the following matrices A

(i) $\begin{pmatrix} 1 & 0 & -1 \\ 2 & 3 & 1 \end{pmatrix}$

(ii) $\begin{pmatrix} 2 & -1 & 0 & 1 \\ 1 & 1 & -1 & 1 \\ -1 & 1 & 2 & 0 \end{pmatrix}$

(iii) $\begin{pmatrix} 1 & 0 & -1 \\ 2 & 1 & 2 \\ -1 & 3 & 1 \\ -1 & 0 & 1 \end{pmatrix}$

(iv) $\begin{pmatrix} 3 & -1 & 0 & 1 & 1 \\ -2 & 0 & 2 & 1 & -1 \\ 1 & 2 & 0 & 0 & 1 \\ -1 & 1 & -2 & -2 & 0 \end{pmatrix}$

2. Determine which of the following matrices are equivalent

(i) $\begin{pmatrix} 1 & -1 & 1 & 2 \\ 0 & 1 & 2 & 0 \\ 1 & 0 & 3 & 2 \end{pmatrix}$ (ii) $\begin{pmatrix} 0 & 1 & 0 & -2 \\ 0 & -1 & 0 & 2 \\ 0 & 2 & 0 & -4 \end{pmatrix}$

(iii) $\begin{pmatrix} 2 & -1 & 0 & 1 \\ -1 & 1 & 1 & -2 \\ 1 & 1 & 3 & -4 \end{pmatrix}$ (iv) $\begin{pmatrix} -1 & 1 & 0 & 1 \\ 2 & 0 & 1 & 1 \\ 1 & 1 & 2 & -1 \end{pmatrix}$

(v) $\begin{pmatrix} 1 & 0 & -1 & 1 \\ -1 & 0 & 2 & 1 \\ 0 & 0 & 1 & 1 \end{pmatrix}$

37

CHAPTER 2

Determinants

Let $M_n(K)$ denote the set of all $n \times n$ matrices over a field K.

2.1 2 × 2 and 3 × 3 Determinants

We shall first motivate our discussion of $n \times n$ determinants by considering 2 × 2 and 3 × 3 determinants.

Let $A = \begin{pmatrix} \alpha_{11} & \alpha_{12} \\ \alpha_{21} & \alpha_{22} \end{pmatrix} \in M_2(K)$, then define det : $M_2(K) \to K$ by

$$\det A = \alpha_{11}\alpha_{22} - \alpha_{12}\alpha_{21} \in K$$

The determinant function arises naturally in the solution of linear equations, for consider the system of two linear equations in two variables

$$\alpha_{11}x_1 + \alpha_{12}x_2 = \beta_1$$
$$\alpha_{21}x_1 + \alpha_{22}x_2 = \beta_2$$

then

$$x_1 = \frac{\alpha_{22}\beta_1 - \alpha_{12}\beta_2}{\alpha_{11}\alpha_{22} - \alpha_{12}\alpha_{21}} = \frac{\det \begin{pmatrix} \beta_1 & \alpha_{12} \\ \beta_2 & \alpha_{22} \end{pmatrix}}{\det A}$$

and

$$x_2 = \frac{\alpha_{21}\beta_1 - \alpha_{11}\beta_2}{\alpha_{12}\alpha_{21} - \alpha_{11}\alpha_{22}} = \frac{\det \begin{pmatrix} \alpha_{11} & \beta_1 \\ \alpha_{21} & \beta_2 \end{pmatrix}}{\det A}$$

The following properties of the determinant function are easily verified

(i) $\det \begin{pmatrix} \alpha_{11} & \alpha_{12} \\ \alpha_{21} + \beta_{21} & \alpha_{22} + \beta_{22} \end{pmatrix} = \det \begin{pmatrix} \alpha_{11} & \alpha_{12} \\ \alpha_{21} & \alpha_{22} \end{pmatrix} + \det \begin{pmatrix} \alpha_{11} & \alpha_{12} \\ \beta_{21} & \beta_{22} \end{pmatrix}$

i.e. $\alpha_{11}(\alpha_{22} + \beta_{22}) - \alpha_{12}(\alpha_{21} + \beta_{21}) = (\alpha_{11}\alpha_{22} - \alpha_{12}\alpha_{21})$
$+ (\alpha_{11}\beta_{22} - \alpha_{12}\beta_{21})$. In other words, if $r_1 = (\alpha_{11}, \alpha_{12})$, $r_2 = (\alpha_{21}, \alpha_{22})$, $r_2' = (\beta_{21}, \beta_{22})$, then

$$\det \begin{pmatrix} r_1 \\ r_2 + r_2' \end{pmatrix} = \det \begin{pmatrix} r_1 \\ r_2 \end{pmatrix} + \det \begin{pmatrix} r_1 \\ r_2' \end{pmatrix}$$

Similarly, the following are also easily verified

(ii) $\det \begin{pmatrix} \alpha\, r_1 \\ r_2 \end{pmatrix} = \alpha \det \begin{pmatrix} r_1 \\ r_2 \end{pmatrix}$, for all $\alpha \in K$,

(iii) $\det \begin{pmatrix} r_1 + \alpha\, r_2 \\ r_2 \end{pmatrix} = \det \begin{pmatrix} r_1 \\ r_2 \end{pmatrix}$, for all $\alpha \in K$

(iv) $\det \begin{pmatrix} r_1 \\ r_1 \end{pmatrix} = 0$

(v) $\det \begin{pmatrix} r_1 \\ r_2 \end{pmatrix} = -\det \begin{pmatrix} r_2 \\ r_1 \end{pmatrix}$

(vi) $\det \begin{pmatrix} 1 & 0 \\ 0 & 1 \end{pmatrix} = 1$

(vii) $\det A^t = \det A$.

(i) and (ii) imply that the determinant function is linear on the rows of A, that is

$$\det \begin{pmatrix} r_1 + \alpha\, r_1' \\ r_2 \end{pmatrix} = \det \begin{pmatrix} r_1 \\ r_2 \end{pmatrix} + \alpha \det \begin{pmatrix} r_1' \\ r_2 \end{pmatrix}$$

(vii) implies that statements (i)-(v) can also be stated for the columns of the matrix A, for example

$$\det (c_1 + \alpha\, c_1', c_2) = \det (c_1, c_2) + \alpha \det (c_1', c_2)$$

Not all of the above results are independent as will be seen later when we consider the general case. The above properties are also useful in evaluating determinants, although it should be emphasized that for 2×2 matrices the evaluation can be performed easily by applying the definition directly, as for example in

$$\det \begin{pmatrix} x & y \\ x^3 & y^3 \end{pmatrix} = xy^3 - x^3 y = xy(y^2 - x^2).$$

This determinant may also be evaluated by applying the above rules,

$$\det \begin{pmatrix} x & y \\ x^3 & y^3 \end{pmatrix} = xy \det \begin{pmatrix} 1 & 1 \\ x^2 & y^2 \end{pmatrix} \qquad \text{by (ii)}$$

$$= xy \det \begin{pmatrix} 1 & 0 \\ x^2 & y^2 - x^2 \end{pmatrix} \qquad \text{by (iii)}$$

$$= xy\,(y^2 - x^2)$$

We now consider the case of 3×3 determinants.

Let $A = \begin{pmatrix} \alpha_{11} & \alpha_{12} & \alpha_{13} \\ \alpha_{21} & \alpha_{22} & \alpha_{23} \\ \alpha_{31} & \alpha_{32} & \alpha_{33} \end{pmatrix} \in M_3(K)$, then define $\det : M_3(K) \to K$

by

$$\det A = \alpha_{11}\alpha_{22}\alpha_{33} + \alpha_{12}\alpha_{23}\alpha_{31} + \alpha_{13}\alpha_{21}\alpha_{32} - \alpha_{11}\alpha_{23}\alpha_{32}$$
$$- \alpha_{12}\alpha_{21}\alpha_{33} - \alpha_{13}\alpha_{22}\alpha_{31}$$

Although there are ways of memorizing this formula, it is not pleasant to use for the actual evaluation of a determinant. But again the properties (i)-(vii) suitably adapted, are valid in this case.

In addition, if we let A_{ij} be the 2×2 matrix obtained by deleting the ith row and jth column of A $(i, j = 1, 2, 3)$ then

$$\begin{aligned} \text{(viii)} \quad \det A &= \alpha_{11} \det A_{11} - \alpha_{12} \det A_{12} + \alpha_{13} \det A_{13} \\ &= -\alpha_{21} \det A_{21} + \alpha_{22} \det A_{22} - \alpha_{23} \det A_{23} \\ &= \alpha_{31} \det A_{31} - \alpha_{32} \det A_{32} + \alpha_{33} \det A_{33} \\ &= \alpha_{11} \det A_{11} - \alpha_{21} \det A_{21} + \alpha_{31} \det A_{31} \\ &= -\alpha_{12} \det A_{12} + \alpha_{22} \det A_{22} - \alpha_{32} \det A_{32} \\ &= \alpha_{13} \det A_{13} - \alpha_{23} \det A_{23} + \alpha_{33} \det A_{33}. \end{aligned}$$

These equations are verified by expanding the right hand side in each case, for example

$$\alpha_{11} \det A_{11} - \alpha_{12} \det A_{12} + \alpha_{13} \det A_{13}$$

$$= \alpha_{11} \det \begin{pmatrix} \alpha_{22} & \alpha_{23} \\ \alpha_{32} & \alpha_{33} \end{pmatrix} - \alpha_{12} \det \begin{pmatrix} \alpha_{21} & \alpha_{23} \\ \alpha_{31} & \alpha_{33} \end{pmatrix} + \alpha_{13} \det \begin{pmatrix} \alpha_{21} & \alpha_{22} \\ \alpha_{31} & \alpha_{32} \end{pmatrix}$$

$$= \alpha_{11}(\alpha_{22}\alpha_{33} - \alpha_{23}\alpha_{32}) - \alpha_{12}(\alpha_{21}\alpha_{33} - \alpha_{23}\alpha_{31}) + \alpha_{13}(\alpha_{21}\alpha_{32} - \alpha_{22}\alpha_{31})$$

$$= \det A$$

To illustrate the above, we evaluate three examples.

EXAMPLES

1.
$$\det \begin{pmatrix} 2 & -1 & 3 \\ 1 & 0 & 2 \\ -3 & 2 & 1 \end{pmatrix} = \det \begin{pmatrix} 0 & -1 & -1 \\ 1 & 0 & 2 \\ 0 & 2 & 7 \end{pmatrix},$$ by subtracting 2 × row 2 from row 1 and adding 3 × row 2 to row 3 and using (iii)

$$= -\det \begin{pmatrix} -1 & -1 \\ 0 & 5 \end{pmatrix} \quad \text{by (viii)}$$

$$= 5$$

2.
$$\det \begin{pmatrix} 1 & 1 & 1 \\ x & y & z \\ x^2 & y^2 & z^2 \end{pmatrix} = \det \begin{pmatrix} 1 & 0 & 0 \\ x & y-x & z-x \\ x^2 & y^2-x^2 & z^2-x^2 \end{pmatrix} \quad \text{by (iii)}$$

$$= \det \begin{pmatrix} y-x & z-x \\ y^2-x^2 & z^2-x^2 \end{pmatrix}, \quad \text{by (viii)}$$

$$= (y-x)(z-x) \det \begin{pmatrix} 1 & 1 \\ y+x & z+x \end{pmatrix}, \text{by (ii)}$$

$$= (y-x)(z-x) \det \begin{pmatrix} 1 & 0 \\ y+x & z-y \end{pmatrix}, \text{by (iii)}$$

$$= (y-x)(z-x)(z-y)$$

3.
$$\det \begin{pmatrix} y+z & x & x^3 \\ z+x & y & y^3 \\ x+y & z & z^3 \end{pmatrix} = \det \begin{pmatrix} x+y+z & x & x^3 \\ x+y+z & y & y^3 \\ x+y+z & z & z^3 \end{pmatrix}$$

$$= (x+y+z) \det \begin{pmatrix} 1 & x & x^3 \\ 0 & y-x & y^3-x^3 \\ 0 & z-x & z^3-x^3 \end{pmatrix}$$

$$= (x+y+z)(y-x)(z-x) \det \begin{pmatrix} 1 & y^2+yx+x^2 \\ 1 & z^2+zx+x^2 \end{pmatrix}$$

$$= (x+y+z)(y-x)(z-x) \det \begin{pmatrix} 1 & y^2+yx+x^2 \\ 0 & z^2-y^2+(z-y)x \end{pmatrix}$$

$$= (x+y+z)^2 (y-x)(z-x)(z-y)$$

41

Exercises 2.1

In the following exercises we write $|A|$ for det A.

1. Evaluate the following determinants

(i) $\begin{vmatrix} 3 & 2 \\ 4 & 5 \end{vmatrix}$ (ii) $\begin{vmatrix} 2 & 0 & -1 \\ 1 & 2 & 2 \\ 3 & 2 & 4 \end{vmatrix}$ (iii) $\begin{vmatrix} 2 & 1 & 2 \\ -1 & 1 & 5 \\ -1 & 2 & 3 \end{vmatrix}$

(iv) $\begin{vmatrix} -2 & 1 & 0 \\ 5 & 0 & 1 \\ 0 & 2 & 2 \end{vmatrix}$ (v) $\begin{vmatrix} 2 & 3 & 1 \\ 1 & 0 & 1 \\ 3 & 3 & 2 \end{vmatrix}$

2. Evaluate the following determinants in factors

(i) $\begin{vmatrix} a & b+c & a^2 \\ b & c+a & b^2 \\ c & a+b & c^2 \end{vmatrix}$ (ii) $\begin{vmatrix} a^2 & (b+c)^2 & bc \\ b^2 & (c+a)^2 & ca \\ c^2 & (a+b)^2 & ab \end{vmatrix}$

(iii) $\begin{vmatrix} b^2+c^2 & (b-c)^2 & a^2+2bc \\ c^2+a^2 & (c-a)^2 & b^2+2ca \\ a^2+b^2 & (a-b)^2 & c^2+2ab \end{vmatrix}$

(iv) $\begin{vmatrix} (x-a)^2 & (x+a)^2 & 2y^2+2z^2 \\ (y-a)^2 & (y+a)^2 & 2z^2+2x^2 \\ (z-a)^2 & (z+a)^2 & 2x^2+2y^2 \end{vmatrix}$

(v) $\begin{vmatrix} a(b+c) & (b+c)^2 & a^2-2bc \\ b(c+a) & (c+a)^2 & b^2-2ca \\ c(a+b) & (a+b)^2 & c^2-2ab \end{vmatrix}$

3. Find the general value of θ which satisfies the equation

$$\begin{vmatrix} 1+\sin^2\theta & \cos^2\theta & 4\sin 2\theta \\ \sin^2\theta & 1+\cos^2\theta & 4\sin 2\theta \\ \sin^2\theta & \cos^2\theta & 1+4\sin 2\theta \end{vmatrix} = 0$$

42

4. Solve the equation

$$\begin{vmatrix} a-x & b-x & c \\ a-x & c & b-x \\ a & b-x & c-x \end{vmatrix} = 0$$

2.2 $n \times n$ Determinants

In the previous section, we saw that even for the case $n = 3$, the explicit expression for the determinant of a matrix is very complicated. It was shown that in the evaluation of determinants the explicit form was not important, for the expansion of determinants the properties of determinants were exploited. For higher values of n, we seek a function which has these desirable properties. In this section, we show that such a function exists and is unique. Our main result will be to prove

THEOREM 2.1   *If* $A = (\alpha_{ij}) = (r_1, r_2, \ldots, r_n) \in M_n(K)$, *where* $r_i = (\alpha_{i1}, \alpha_{i2}, \ldots, \alpha_{in})$ $(i = 1, \ldots, n)$ *are the rows of A, then there exists a function* $D : M_n(K) \to K$ *such that*
    (i) *D is linear on the rows of A, i.e. for* $i = 1, 2, \ldots, n$
$$D(r_1, \ldots, \alpha r_i + r_i', \ldots, r_n) = \alpha D(r_1, \ldots, r_i, \ldots, r_n)$$
$$+ D(r_1, \ldots, r_i', \ldots, r_n),$$
    (ii) *if two adjacent rows of A are equal, i.e.* $r_k = r_{k+1}$ *for some k such that* $k = 1, 2, \ldots, n-1$, *then* $D(A) = 0$,
    (iii) $D(I_n) = 1$, *where* I_n *is the identity* $n \times n$ *matrix. D is uniquely determined by the above properties* (i), (ii) *and* (iii) *and is called the* **determinant function.**
The reader should note that the r_i are **rows** of the matrix A, which has been written in the form $A = (r_1, \ldots, r_n)$ for obvious typographical reasons.

Before proving the theorem, we show that the additional properties of determinants which have already been verified for $n = 2, 3$ are consequences of the three defining properties (i), (ii) and (iii). Furthermore, we shall also denote the determinant function by det and the value by det A or the more traditional $|A|$ or $|\alpha_{ij}|$ depending on what is most convenient in the context.

PROPOSITION 2.2   *For* $j = 1, 2, \ldots, n-1$, *if* $B = (r_1, \ldots, r_{j+1}, r_j, \ldots, r_n)$, *then det* $B = -\det A$, *i.e. if two adjacent rows of A are interchanged, the value of the determinant is changed by a factor of* -1.

PROOF

$$0 = \det(r_1, \ldots, r_j + r_{j+1}, r_j + r_{j+1}, \ldots, r_n) \qquad \text{by (ii)}$$

$$
\begin{aligned}
= \ & \det(r_1, \ldots, r_j, r_j, \ldots, r_n) \\
& + \det(r_1, \ldots, r_j, r_{j+1}, \ldots, r_n) \\
& + \det(r_1, \ldots, r_{j+1}, r_j, \ldots, r_n) \\
& + \det(r_1, \ldots, r_{j+1}, r_{j+1}, \ldots, r_n),
\end{aligned}
$$

by repeated application of (i)

$$= 0 + \det A + \det B + 0, \qquad \text{by (ii)},$$

that is

$$\det B = -\det A. \qquad \blacksquare$$

PROPOSITION 2.3 *If two rows in A are equal, then*

$$det\ A = 0.$$

PROOF Suppose that the ith and jth rows of A are equal. Interchange the jth row with the succeeding rows adjacent to it until it is adjacent to the ith row. By Proposition 2.2, the sign of the determinant changes at every interchange, but now we have a determinant with adjacent rows equal and by Theorem 2.1 (ii), $\det A = 0$. $\qquad \blacksquare$

PROPOSITION 2.4 *If $j \neq i$, then*
$$det(r_1, \ldots, r_i + \alpha r_j, \ldots, r_j, \ldots, r_n) = \det(r_1, \ldots, r_i, \ldots, r_n),$$
i.e. if a scalar multiple of one row is added to another row then the value of the determinant is unaltered.

PROOF $\det(r_1, \ldots, r_i + \alpha r_j, \ldots, r_j, \ldots, r_n)$

$$
\begin{aligned}
= \ & \det(r_1, \ldots, r_i, \ldots, r_j, \ldots, r_n) \\
& + \alpha \det(r_1, \ldots, r_j, \ldots, r_j, \ldots, r_n) \quad \text{by Theorem 2.1(i)} \\
= \ & \det(r_1, \ldots, r_i, \ldots, r_n) \qquad \text{by Proposition 2.3} \qquad \blacksquare
\end{aligned}
$$

In order to prove the theorem we need the following

DEFINITION 2.5 *If $A = (\alpha_{ij}) \in M_n(K)$, let $A_{ij} \in M_{n-1}(K)$ be the $(n-1) \times (n-1)$ matrix obtained by deleting the ith row and jth column of A. A_{ij} is called the **minor** of the element α_{ij}. The **cofactor** of α_{ij} in A is*

$$c_{ij} = (-1)^{i+j} \det A_{ij} \in K$$

EXAMPLE

If $A = \begin{pmatrix} 1 & -1 & 1 \\ 2 & 0 & -1 \\ 1 & 1 & 2 \end{pmatrix}$, then

$$c_{12} = (-1)^3 \det \begin{pmatrix} 2 & -1 \\ 1 & 2 \end{pmatrix} = -5$$

$$c_{23} = (-1)^5 \det \begin{pmatrix} 1 & -1 \\ 1 & 1 \end{pmatrix} = -2$$

We note that the sign $(-1)^{i+j}$ is given by the pattern

$$\begin{pmatrix} + & - & + & - & . & . & . \\ - & + & - & + & . & . & . \\ + & - & + & - & . & . & . \\ . & . & . & & & & \\ . & . & . & & & & \\ . & . & . & & & & \end{pmatrix}$$

We are now in a position to give the

PROOF OF THEOREM 2.1   The proof is divided into two parts. We first prove the existence of the function D and then show that it is unique.

EXISTENCE   For a fixed $j = 1, 2, \ldots, n$, define $D : M_n(K) \to K$ by

$$D(A) = \sum_{i=1}^{n} \alpha_{ij} c_{ij}$$

The proof is by induction on n, that is, we assume that we have been able to define determinants for all integers $< n$.
If $n = 1$, then the determinant function exists, i.e. $D(\alpha) = \alpha, \alpha \in K$ (we have also verified this for $n = 2, 3$).

We now show that D satisfies the three properties (i), (ii) and (iii) in the statement of the theorem.

(i) Consider D as a function of the kth row of A, and consider for a fixed $1 \leqslant i \leqslant n$, the term

$$\alpha_{ij} c_{ij} = (-1)^{i+j} \alpha_{ij} \det (A_{ij})$$

If $i \neq k$, α_{ij} does not depend on the kth row and by induction, since A_{ij} contains elements from the kth row of A, det A_{ij} depends linearly on the kth row of A. If $i = k$, α_{ij} depends linearly on the kth row of A and det A_{ij} does not depend on the kth row of A.
Thus, each term in $D(A)$ depends linearly on the kth row of A and so also does $D(A)$.

45

(ii) Suppose two adjacent rows of A are equal, say $r_k = r_{k+1}$ for some $k = 1, 2, \ldots, n-1$. Let $i \neq k, k+1$, then A_{ij} has two adjacent rows equal and by the induction hypothesis $c_{ij} = (-1)^{i+j} \det(A_{ij}) = 0$. Thus, we have

$$\det A = \alpha_{kj}(-1)^{j+k} \det(A_{kj}) + \alpha_{k+1,j}(-1)^{j+k+1} \det(A_{k+1,j})$$

But $A_{kj} = A_{k+1,j}$ and so $\det(A_{kj}) = \det(A_{k+1,j})$ and $\alpha_{kj} = \alpha_{k+1,j}$, thus we have

$$\det A = \alpha_{kj}(-1)^{j+k} \det(A_{kj})(1-1) = 0.$$

(iii) If $A = I_n$, the only term which gives a non-zero contribution in $D(A)$ is

$$(-1)^{i+i} \alpha_{ii} \det(A_{ii}) = 1 \det I_{n-1} = 1$$

by the induction hypothesis, i.e.

$$D(I_n) = 1$$

UNIQUENESS   Let D, D' : $M_n(K) \rightarrow K$ satisfy the three conditions (i), (ii) and (iii), then we show that

$$D(r_1, r_2, \ldots, r_n) = D'(r_1, r_2, \ldots, r_n)$$

for all $A = (r_1, r_2, \ldots, r_n) \in M_n(K)$. If $A = (r_1, r_2, \ldots, r_n)$ and

$$\Delta(r_1, r_2, \ldots, r_n) = D(r_1, r_2, \ldots, r_n) - D'(r_1, r_2, \ldots, r_n),$$

then we must show that

$$\Delta(r_1, r_2, \ldots, r_n) = 0$$

It is easily verified that
(a) Δ is linear in the rows of A,
(b) if two rows of A are equal then $\Delta(r_1, r_2, \ldots, r_n) = 0$,
(c) if $(j_1, j_2, \ldots, j_n)$ is a rearrangement of $(1, 2, \ldots, n)$ then

$$\Delta(r_{j_1}, r_{j_2}, \ldots, r_{j_n}) = \pm \Delta(r_1, r_2, \ldots, r_n)$$

(d) $\Delta(I_n) = 0$.
Now, let

$$e_1 = (1, 0, \ldots, 0), \ e_2 = (0, 1, 0, \ldots, 0), \ldots,$$
$$e_n = (0, 0, \ldots, 0, 1),$$

then for $i = 1, 2, \ldots, n$

$$r_i = (\alpha_{i1}, \alpha_{i2}, \ldots, \alpha_{in}) = \sum_{j=1}^{n} \alpha_{ij} e_j$$

Thus, we have by repeatedly applying (a) above that

$$\Delta(r_1, r_2, \ldots, r_n)$$

$$= \Delta \left(\sum_{k=1}^{n} \alpha_{1k} e_k, \ldots, \sum_{k=1}^{n} \alpha_{nk} e_k \right)$$

$$= \sum_{k_1, k_2, \ldots, k_n = 1}^{n} \alpha_{1k_1} \alpha_{2k_2} \cdots \alpha_{nk_n} \Delta(e_{k_1}, e_{k_2}, \ldots, e_{k_n})$$

where the sum ranges over the n^n terms obtained by allowing
$k_1, k_2, \ldots, k_n$ to take values between 1 and n.
But, by (b), if any of the e_{k_i}'s are equal, $\Delta(e_{k_1}, e_{k_2}, \ldots, e_{k_n}) = 0$. Thus
it follows that

$$\Delta(r_1, r_2, \ldots, r_n)$$

$$= \sum_{(k_1, k_2, \ldots, k_n)} \alpha_{1k_1} \alpha_{2k_2} \cdots \alpha_{nk_n} \Delta(e_{k_1}, e_{k_2}, \ldots, e_{k_n})$$

where the sum ranges over all rearrangements $(k_1, k_2, \ldots, k_n)$ of
$(1, 2, \ldots, n)$. By (c) and (d), it now follows that if $(k_1, k_2, \ldots, k_n)$ is
a rearrangement of $(1, 2, \ldots, n)$ then

$$\Delta(e_{k_1}, e_{k_2}, \ldots, e_{k_n}) = \pm \Delta(e_1, e_2, \ldots, e_n)$$

$$= 0$$

and thus

$$\Delta(r_1, r_2, \ldots, r_n) = 0 \text{ as required} \qquad \blacksquare$$

Remark 1   By a similar argument to the above, it can be shown that

$$D(r_1, r_2, \ldots, r_n) = \sum \alpha_{1k_1} \alpha_{2k_2} \cdots \alpha_{nk_n} D(e_{k_1}, e_{k_2}, \ldots, e_{k_n})$$

$$= \sum \text{sgn}(k_1, k_2, \ldots, k_n) \alpha_{1k_1} \alpha_{2k_2} \cdots \alpha_{nk_n},$$

where the sum ranges over the $n!$ rearrangements $(k_1, k_2, \ldots, k_n)$ of
$(1, 2, \ldots, n)$ and $\text{sgn}(k_1, k_2, \ldots, k_n) = (-1)^p$, where p is the number
of row interchanges required to obtain the identity matrix I_n from the
matrix $(e_{k_1}, e_{k_2}, \ldots, e_{k_n})$. This formula is often given as the definition
of the determinant of a matrix; the reader should verify this formula in
the cases $n = 2, 3$.

Remark 2   From the definition of D, it follows that if a matrix A has
a column of zeros then $D(A) = 0$. Furthermore, if A has a row of zeros,
it can be proved that $D(A) = 0$ (see Exercise 2.2, No. 9).

Remark 3 The uniqueness of the determinant function implies that the same value for $D(A)$ is obtained whatever value for j such that $1 \leqslant j \leqslant n$ has been used in the definition of $D(A)$. We call this the **expansion of the determinant of A in terms of the jth column of A**.

Similarly, for each $1 \leqslant i \leqslant n$

$$D(A) = \sum_{j=1}^{n} \alpha_{ij} c_{ij}$$

gives the **expansion of the determinant in terms of the ith row of A** (see p. 56 for a proof of this). Thus, a determinant is expanded in terms of row or column which is most appropriate as illustrated in the examples below.

EXAMPLES

1. $\begin{vmatrix} -1 & 0 & 0 & 3 \\ 1 & 3 & 4 & 1 \\ 2 & 2 & 1 & 5 \\ 0 & 1 & 2 & 7 \end{vmatrix} = \begin{vmatrix} -1 & 0 & 0 & 3 \\ 0 & 3 & 4 & 4 \\ 0 & 2 & 1 & 11 \\ 0 & 1 & 2 & 7 \end{vmatrix}$ Using Proposition 2.4

$= - \begin{vmatrix} 3 & 4 & 4 \\ 2 & 1 & 11 \\ 1 & 2 & 7 \end{vmatrix}$ Using the definition of determinants

$= - \begin{vmatrix} 0 & -2 & -17 \\ 0 & -3 & -3 \\ 1 & 2 & 7 \end{vmatrix}$

$= - \begin{vmatrix} -2 & -17 \\ -3 & -3 \end{vmatrix}$

$= 45$

2. $\begin{vmatrix} 1 & x & y & 1 \\ 1 & x & x & x \\ x & 1 & xy & y \\ x & x & xy & 1 \end{vmatrix} = \begin{vmatrix} 1 & x & y & 1 \\ 0 & 0 & x-y & x-1 \\ 0 & 1-x^2 & 0 & y-x \\ 0 & x-x^2 & 0 & 1-x \end{vmatrix}$

$$= \begin{vmatrix} 0 & x-y & x-1 \\ 1-x^2 & 0 & y-x \\ x-x^2 & 0 & 1-x \end{vmatrix}$$

$$= -(x-y) \begin{vmatrix} 1-x^2 & y-x \\ x-x^2 & 1-x \end{vmatrix}$$

$$= -(x-y)(1-x) \begin{vmatrix} 1-x^2 & y-x \\ x & 1 \end{vmatrix}$$

$$= -(x-y)(1-x) \begin{vmatrix} 1 & y \\ x & 1 \end{vmatrix}$$

$$= (x-1)(x-y)(1-xy)$$

3. Let

$$D_n = \begin{vmatrix} 1 & 1 & \ldots & 1 \\ x_1 & x_2 & \ldots & x_n \\ x_1^2 & x_2^2 & \ldots & x_n^2 \\ \cdot & \cdot & & \cdot \\ \cdot & \cdot & & \cdot \\ \cdot & \cdot & & \cdot \\ x_1^{n-1} & x_2^{n-1} & \ldots & x_n^{n-1} \end{vmatrix}$$

then we prove that

$$D_n = \prod_{n \geqslant i > j \geqslant 1} (x_i - x_j)$$

The proof is by induction on n, the result being clearly true when $n = 2$, i.e.

$$D_2 = \begin{vmatrix} 1 & 1 \\ x_1 & x_2 \end{vmatrix} = x_2 - x_1$$

If r_i ($i = 1, \ldots, n$) denotes the ith row of D_n, then carrying out successively the following elementary row operations
$r_n \rightarrow r_n - x_1 r_{n-1}, r_{n-1} \rightarrow r_{n-1} - x_1 r_{n-2}, \ldots, r_2 \rightarrow r_2 - x_1 r_1$ we obtain

$$D_n = \begin{vmatrix} 1 & 1 & & \ldots & 1 \\ 0 & x_2 - x_1 & & \ldots & x_n - x_1 \\ 0 & x_2(x_2 - x_1) & & \ldots & x_n(x_n - x_1) \\ \cdot & \cdot & & & \cdot \\ \cdot & \cdot & & & \cdot \\ 0 & x_2^{n-2}(x_2 - x_1) & & \ldots & x_n^{n-2}(x_n - x_1) \end{vmatrix}$$

$$= (x_2 - x_1)(x_3 - x_1) \ldots (x_n - x_1) D_{n-1}$$

The proof is now completed by induction on n. D_n is called the **Vandermonde Determinant**.

4.

Let

$$D_n = \begin{vmatrix} 2 & 1 & 0 & 0 & \ldots & 0 \\ 1 & 2 & 1 & 0 & \ldots & 0 \\ 0 & 1 & 2 & 1 & \ldots & 0 \\ \cdot & \cdot & \cdot & \cdot & & \cdot \\ \cdot & \cdot & \cdot & \cdot & & \cdot \\ \cdot & \cdot & \cdot & \cdot & & \cdot \\ 0 & 0 & 0 & 0 & \ldots 1 & 2 \end{vmatrix}$$

$$= 2D_{n-1} - D_{n-2}$$

Thus we have

$$D_n - D_{n-1} = D_{n-1} - D_{n-2} = D_{n-2} - D_{n-3} = \ldots$$

from which we infer that

$$D_n - D_{n-1} = D_2 - D_1 = 3 - 2 = 1$$

Similarly, we have

$$D_{n-1} - D_{n-2} = 1$$
$$\cdot$$
$$\cdot$$
$$D_2 - D_1 = 1$$

and adding gives

$$D_n = (n-1) + D_1 = n + 1$$

Exercises 2.2

1. Evaluate the determinants

50

(i)
$$\begin{vmatrix} 0 & -1 & 2 & 1 \\ -4 & 3 & -3 & 5 \\ 1 & 0 & 0 & -1 \\ -1 & 1 & 0 & 1 \end{vmatrix}$$

(ii)
$$\begin{vmatrix} 2 & 1 & 0 & 1 \\ -1 & 2 & 1 & 0 \\ 1 & 2 & 5 & 3 \\ 0 & -1 & 1 & 1 \end{vmatrix}$$

(iii)
$$\begin{vmatrix} 2 & -1 & 1 & 1 \\ 5 & 6 & 1 & 2 \\ 1 & -2 & 1 & 1 \\ 3 & -2 & 1 & 1 \end{vmatrix}$$

2. For what values of x is the determinant

$$\begin{vmatrix} x & a & b & c \\ a & x & b & c \\ a & b & x & c \\ a & b & c & x \end{vmatrix} = 0?$$

3. Show that the value of the determinant

$$\begin{vmatrix} 0 & 1 & 1 & 1 \\ 1 & 0 & a+b & a+c \\ 1 & b+a & 0 & b+c \\ 1 & c+a & c+b & 0 \end{vmatrix} = -4(ab+bc+ca)$$

and hence find the values of the determinant when
   (i) a, b and c are the distinct cube roots of unity.
   (ii) a, b and c are distinct fourth roots of unity.

4. Show that the following n-rowed determinants

(i)
$$\begin{vmatrix} 1 & 1 & 1 & \ldots & 1 & 1 \\ -1 & 1 & 1 & \ldots & 1 & 1 \\ 0 & -1 & 1 & \ldots & 1 & 1 \\ \cdot & \cdot & \cdot & & 1 & 1 \\ \cdot & \cdot & \cdot & & & \\ 0 & 0 & 0 & \ldots & -1 & 1 \end{vmatrix} = 2^{n-1}$$

(ii) $$\begin{vmatrix} x & 1 & 1 & \ldots & 1 & 1 \\ 1 & x & 0 & \ldots & 0 & 0 \\ 1 & 1 & x & \ldots & 0 & 0 \\ \cdot & \cdot & \cdot & & & \cdot \\ \cdot & \cdot & \cdot & & & \cdot \\ \cdot & \cdot & \cdot & & & \cdot \\ 1 & 1 & 1 & \ldots & 1 & x \end{vmatrix} = (x-1)\{(x-1)^{n-2} + x^{n-1}\}$$

(iii) $$\begin{vmatrix} 1+x^2 & x & 0 & \ldots & 0 \\ x & 1+x^2 & x & \ldots & 0 \\ 0 & x & 1+x^2 & \ldots & 0 \\ \cdot & \cdot & \cdot & & \cdot \\ \cdot & \cdot & \cdot & & \cdot \\ \cdot & \cdot & \cdot & & \cdot \\ 0 & 0 & 0 & x & 1+x^2 \end{vmatrix} = 1+x^2+\ldots+x^{2n}$$

(iv) $$\begin{vmatrix} a+b & a & a & \ldots & a \\ a & a+b & a & \ldots & a \\ \cdot & \cdot & \cdot & & \cdot \\ \cdot & \cdot & \cdot & & \cdot \\ \cdot & \cdot & \cdot & & \cdot \\ a & a & a & \ldots & a+b \end{vmatrix} = b^{n-1}(na+b)$$

5. If $A = (\alpha_{ij})$ is an $n \times n$ matrix with $\alpha_{11} \neq 0$ and also $\alpha_{ij} \neq 0$ when $i + j > n + 1$, the remaining elements all being zero, find an expression for det A.

6. Let A_n be the $n \times n$ matrix

$$\begin{pmatrix} -2 & 4 & 0 & 0 & \ldots & 0 & 0 \\ 1 & -2 & 4 & 0 & \ldots & 0 & 0 \\ 0 & 1 & -2 & 4 & \ldots & 0 & 0 \\ \cdot & \cdot & \cdot & \cdot & & \cdot & \cdot \\ \cdot & \cdot & \cdot & \cdot & & \cdot & \cdot \\ \cdot & \cdot & \cdot & \cdot & & \cdot & \cdot \\ 0 & 0 & 0 & 0 & 1 & -2 & 4 \\ 0 & 0 & 0 & 0 & \ldots & 1 & -2 \end{pmatrix}$$

Show that det $A_n = 0$ when $n = 3k - 1$ and find det A_n when $n = 3k$.

7. Let $D : M_n(\mathbf{R}) \to \mathbf{R}$ be such that

$$D(A)\, D(B) = D(AB)$$

for all $A, B \in M_n(\mathbf{R})$. Show that either $D(A) = 0$ for all $A \in M_n(\mathbf{R})$ or $D(I_n) = 1$. In the latter case, show that $D(A) \neq 0$ whenever A is invertible.

If $n = 2$, suppose further that $D\begin{pmatrix} 0 & 1 \\ 1 & 0 \end{pmatrix} \neq D\begin{pmatrix} 1 & 0 \\ 0 & 1 \end{pmatrix}$. Prove the following
   (i) $D(0) = 0$,
   (ii) $D(A) = 0$ if $A^2 = 0$,
   (iii) $D(B) = -D(A)$ if B is obtained by interchanging the rows (or columns) of A.

8. If A is an $n \times n$ matrix which has a row of zeros prove that $D(A) = 0$.

9. Let A_n be the $n \times n$ matrix whose (i, j)th element is $|i - j|$. Show that $\det A_n = (-1)^{n-1}(n-1)\, 2^{n-2}$.

2.3 Further Properties of Determinants

In this section, we prove that the determinant function is multiplicative and deduce some important consequences of this result. The proof involves the notion of elementary matrices introduced in Chapter 1. We first prove the following lemma

LEMMA 2.6   *If E is an elementary matrix, then*
   (i) $\det E \neq 0$
   (ii) $\det E^t = \det E$
   (iii) E^{-1} *is an elementary matrix.*

PROOF   The proof is by a case-by-case evaluation of the determinant of the three different types of elementary matrices

$$\det M_i(\alpha) = \det M_i(\alpha)^t = \alpha \qquad (i = 1, \ldots, n)$$
$$\det H_{ij} = \det H_{ij}^t = -1 \qquad (i, j = 1, \ldots, n)$$
$$\det A_{ij}(\alpha) = \det A_{ij}(\alpha)^t = 1 \qquad (i, j = 1, \ldots, n)$$

proving (i) and (ii). To prove (iii), we note that

$$M_i(\alpha)^{-1} = M_i(\alpha^{-1}) \quad H_{ij}^{-1} = H_{ij} \quad A_{ij}(\alpha)^{-1} = A_{ji}(-\alpha) \qquad \blacksquare$$

We now prove our main theorem.

53

THEOREM 2.7   *If E is an elementary n × n matrix, then*

$$\det EA = (\det E)(\det A)$$

for all n × n matrices A.

PROOF   The proof is again by a case-by-case evaluation. Premultiplying a matrix A by $M_i(\alpha)$ is equivalent to multiplying the ith row of A by α, thus

$$\det (M_i(\alpha)A) = \alpha \det A \qquad \text{by Theorem 2.1(i)}$$
$$= (\det M_i(\alpha))(\det A) \quad \text{by Lemma 2.6}$$

Premultiplying a matrix A by H_{ij} is equivalent to interchanging the ith and jth row of A, which by repeated application of Proposition 2.2 gives

$$\det (H_{ij}A) = -\det A = (\det H_{ij})(\det A)$$

Similarly, premultiplying a matrix A by $A_{ij}(\alpha)$ is equivalent to adding α times the jth row of A to the ith row which by Proposition 2.4 and Lemma 2.6 gives

$$\det (A_{ij}(\alpha)A) = \det A = (\det (A_{ij}(\alpha))(\det A) \qquad ▮$$

Our main results are now obtained as corollaries to this theorem.

COROLLARY 1   *A matrix A is non-singular if and only if det A ≠ 0.*

PROOF   If A is non-singular, then by Theorem 1.17 Corollary 2, there exist elementary matrices $E_1, \dots, E_r$ such that

$$E_1 E_2 \dots E_r A = I_n$$

Thus, by the above theorem

$$(\det E_1)(\det E_2) \dots (\det E_r)(\det A) = \det I_n = 1$$

and so $\det A \neq 0$ as required.

Conversely, if A is singular, then by Theorem 1.17 there exist elementary matrices $E_1, \dots, E_r$ such that

$$E_1 E_2 \dots E_r A = R$$

where R is a reduced echelon $n \times n$ matrix which contains a row, and hence a column of zeros. Thus, $\det R = 0$ and since by Lemma 2.6 (i), $\det E_i \neq 0$ ($i = 1, \dots, r$), it follows that $\det A = 0$. ▮

COROLLARY 2   *If A and B are n × n matrices, then*

$$\det AB = (\det A)(\det B)$$

PROOF   If A and B are non-singular matrices, then by Theorem 1.17, there exist elementary matrices $E_1, \ldots, E_r, E'_1, \ldots, E'_s$ such that

$$A = E_1 E_2 \ldots E_r, \quad B = E'_1 E'_2 \ldots E'_s$$

Then, by repeated application of Theorem 2.7, we have

$$\begin{aligned}
\det AB &= \det (E_1 \ldots E_r E'_1 \ldots E'_s) \\
&= (\det E_1) \ldots (\det E_r)(\det E'_1) \ldots (\det E'_s) \\
&= (\det A)(\det B)
\end{aligned}$$

If A is singular, then by Corollary 1, $\det A = 0$ and by Theorem 1.17 there exist elementary matrices $E_1, \ldots, E_r$ such that

$$A = E_1 \ldots E_r R$$

where R contains a row of zeros. Then we have

$$\begin{aligned}
\det (AB) &= (\det E_1) \ldots (\det E_r)(\det RB) \\
&= 0
\end{aligned}$$

since RB also contains a row of zeros.
Thus again

$$(\det AB) = (\det A)(\det B) = 0$$

A similar argument is used if B is singular.   ∎

COROLLARY 3   *The system of n homogeneous equations in n*

variables $\sum\limits_{j=1}^{n} \alpha_{ij} x_j = 0$ $(i = 1, \ldots, n)$ *has a non-trivial solution if and only if* $\det A = 0$, *where* $A = (\alpha_{ij})$.

PROOF   This follows immediately from Corollary 1 and Corollary 2 to Theorem 1.17.   ∎

COROLLARY 4   *If* $A \in M_n(K)$, *then*

$$\det A^t = \det A$$

PROOF   If A is non-singular, then by Corollary 2 to Theorem 1.17

$$A = E_1 \ldots E_r$$

where $E_i\ (i = 1, \ldots, r)$ are elementary matrices. Then using Lemma 2.6 and Theorem 2.7 we have

$$\det A^t = (\det E_r^t) \ldots (\det E_1^t)$$

$$= (\det E_1) \ldots (\det E_r)$$

$$= \det A$$

If A is singular, then by Corollary 1, $\det A = 0$ and

$$A = E_1 \ldots E_r R$$

where $E_1, \ldots, E_r$ are elementary matrices and R is the reduced echelon matrix of A. Since R^t has a row and column of zeros, it follows that $\det R^t = 0$ and thus, we have

$$\det A^t = 0$$

that is,

$$\det A^t = \det A \ (= 0)$$

in this case also. ∎

This last corollary is important in that it shows that the defining properties of a determinant stated in terms of the rows of a matrix A are also true for the columns of A, i.e.

(i) det is linear in the columns of A

(ii) if two adjacent columns of A are equal then $\det A = 0$.

Furthermore, Propositions 2.2, 2.3 and 2.4 can now also be stated for the columns of A and also

$$\sum_{j=1}^{n} \alpha_{ij}\, c_{ij} = \det A \qquad (i = 1, \ldots, n)$$

Of course these statements could have been proved directly for columns only minor modifications of the appropriate proofs being essential.

Exercises 2.3

1. Determine which of the following matrices are non-singular

(i) $\begin{pmatrix} 2 & 1 & 3 \\ 1 & -1 & 2 \\ 4 & 5 & -2 \end{pmatrix}$
(ii) $\begin{pmatrix} -1 & 1 & 1 \\ 2 & 0 & 1 \\ 1 & 3 & 5 \end{pmatrix}$
(iii) $\begin{pmatrix} 1 & 0 & 2 \\ -1 & 1 & 1 \\ 2 & -1 & 1 \end{pmatrix}$

2. If $s_r = \alpha^r + \beta^r + \gamma^r$, by expressing the determinant as the product of two determinants show that

$$\begin{vmatrix} 3 & s_1 & s_2 \\ s_1 & s_2 & s_3 \\ s_2 & s_3 & s_4 \end{vmatrix} = (\alpha - \beta)^2 \, (\alpha - \gamma)^2 \, (\beta - \gamma)^2$$

Evaluate the determinant

$$\begin{vmatrix} 3 & s_1 & s_3 \\ s_1 & s_2 & s_4 \\ s_2 & s_3 & s_5 \end{vmatrix}$$

3. If $s_r = \alpha^r + \beta^r + \gamma^r$, show that

$$\begin{vmatrix} s_r & s_{r+1} & s_{r+2} \\ s_{r+1} & s_{r+2} & s_{r+3} \\ s_{r+2} & s_{r+3} & s_{r+4} \end{vmatrix} = \alpha^r \, \beta^r \, \gamma^r \, (\beta - \gamma)^2 \, (\gamma - \alpha)^2 \, (\alpha - \beta)^2$$

If α, β, γ are roots of $x^3 + bx + c$, evaluate this expression in terms of b and c.

4. Prove that the product of the matrices

$$\begin{pmatrix} a_1 + ib_1 & c_1 + id_1 \\ -(c_1 - id_1) & a_1 - ib_1 \end{pmatrix} \begin{pmatrix} a_2 + ib_2 & c_2 + id_2 \\ -(c_2 - id_2) & a_2 - ib_2 \end{pmatrix}$$

is another matrix of the same form, where $a_1, b_1, c_1, d_1, a_2, b_2, c_2, d_2$ are integers. By evaluating the determinants of these matrices prove that the product of a sum of squares of four integers with a sum of squares of four integers is the sum of squares of four integers.

5. Let A be the matrix

$$\begin{pmatrix} a + bx + cx^2 & a + by + cy^2 & a + bz + cz^2 \\ c + ax + bx^2 & c + ay + by^2 & c + az + bz^2 \\ b + cx + ax^2 & b + cy + ay^2 & b + cz + az^2 \end{pmatrix}$$

and let B be the matrix

$$\begin{pmatrix} 1 & 1 & 1 \\ x & y & z \\ x^2 & y^2 & z^2 \end{pmatrix}$$

Find a matrix C such that $A = CB$.

Show that the determinant of A is equal to

$$\begin{vmatrix} a + b + c & a + b + c & a + b + c \\ ax + cy + bz & bx + ay + cz & cx + by + az \\ ax^2 + cy^2 + bz^2 & bx^2 + ay^2 + cz^2 & cx^2 + by^2 + az^2 \end{vmatrix}$$

6. Determine whether the following systems of homogeneous equations have non-trivial solutions

(i) $\begin{aligned} x + 2y - z &= 0 \\ -2x + y + 2z &= 0 \\ x - y + z &= 0 \end{aligned}$ (ii) $\begin{aligned} x + 2y - z &= 0 \\ -2x + y + 2z &= 0 \\ x - y - z &= 0 \end{aligned}$

7. Show that if $k \neq 1$ there is always a solution to the equations

$$2x - y - z = 6$$
$$x - 2y + z = p$$
$$x + ky - 2z = p^2$$

whatever the values of p and for a fixed p the solution is unique. Find the solution when $k = 2, p = 1$. Show that if $k = 1$ there are exactly two values of p for which the equations have solutions; find these values and solve the equations completely in each case.

8. For which values of k does the following system of linear equations have (i) no solution, (ii) a unique solution and (iii) more than one solution?

$$x + (k + 2)y + 2z = 2$$
$$(k^2 + 1)x + 2(k + 2)y + 4z = 4k$$
$$3x + 9y + 3(k + 1)z = 6k$$

9. Solve Exercises 1.3, Nos. 3-7, with the methods now available.

2.4 The Inverse of a Matrix

In Chapter 1, a method was presented for inverting a matrix. We now give an alternative method which is not as efficient for practical purposes but is more useful theoretically.

By the definition of determinants, we have for $j = 1, 2, \ldots, n$

$$\sum_{i=1}^{n} \alpha_{ij} c_{ij} = \det A$$

Furthermore, for $k \neq j$, we have

$$\sum_{i=1}^{n} \alpha_{ij} c_{ik} = \begin{vmatrix} \alpha_{11} & \cdots & \alpha_{1j} & \cdots & \alpha_{1j} & \cdots & \alpha_{1n} \\ \alpha_{21} & \cdots & \alpha_{2j} & \cdots & \alpha_{2j} & \cdots & \alpha_{2n} \\ \cdot & & \cdot & & \cdot & & \cdot \\ \cdot & & \cdot & & \cdot & & \cdot \\ \cdot & & \cdot & & \cdot & & \cdot \\ \alpha_{n1} & \cdots & \alpha_{nj} & \cdots & \alpha_{nj} & \cdots & \alpha_{nn} \end{vmatrix} = 0$$

Combining these two, we obtain

$$\sum_{i=1}^{n} \alpha_{ij} c_{ik} = \delta_{jk} \det A \tag{1}$$

DEFINITION 2.8 *If $A \in M_n(K)$ then the **adjoint** of A, denoted by adj A is defined by*

$$adj\ A = (c_{ij})^t$$

i.e. the $n \times n$ matrix obtained by placing c_{ji} in the (i, j)-position.

Then, using (1), we have

$$(adj\ A)\ A = (c_{ji})(\alpha_{ij})$$

$$= \left(\sum_{k=1}^{n} c_{ki} \alpha_{kj} \right)$$

$$= \begin{pmatrix} \det A & \cdots & 0 \\ \cdot & \cdot & \cdot \\ \cdot & & \cdot \\ 0 & \cdots & \det A \end{pmatrix}$$

$$= (\det A)\ I_n$$

Thus, if $\det A \neq 0$, then

$$\frac{adj\ A}{\det A}\ A = I_n$$

or in other words, since the inverse of a matrix is unique,

$$A^{-1} = \frac{adj\ A}{\det A}$$

EXAMPLE   Let $A = \begin{pmatrix} 1 & 0 & -1 \\ 2 & 1 & -1 \\ 1 & 2 & 5 \end{pmatrix}$, then $\det A = 4$ and

$$\text{adj } A = \begin{pmatrix} +\begin{vmatrix} 1 & -1 \\ 2 & 5 \end{vmatrix} & -\begin{vmatrix} 0 & -1 \\ 2 & 5 \end{vmatrix} & +\begin{vmatrix} 0 & -1 \\ 1 & -1 \end{vmatrix} \\[2mm] -\begin{vmatrix} 2 & -1 \\ 1 & 5 \end{vmatrix} & +\begin{vmatrix} 1 & -1 \\ 1 & 5 \end{vmatrix} & -\begin{vmatrix} 1 & -1 \\ 2 & -1 \end{vmatrix} \\[2mm] +\begin{vmatrix} 2 & 1 \\ 1 & 2 \end{vmatrix} & -\begin{vmatrix} 1 & 0 \\ 1 & 2 \end{vmatrix} & +\begin{vmatrix} 1 & 0 \\ 2 & 1 \end{vmatrix} \end{pmatrix}$$

Therefore

$$A^{-1} = \tfrac{1}{4} \begin{pmatrix} 7 & -2 & 1 \\ -11 & 6 & -1 \\ 3 & -2 & 1 \end{pmatrix}$$

As mentioned earlier, this is not an efficient method for calculating the inverse for large n, it involves the evaluation of one $n \times n$ determinant and $n^2 (n-1) \times (n-1)$ determinants.

This may be applied to the solution of linear equations. Consider the n linear equations in n variables

$$\sum_{j=1}^{n} \alpha_{ij} x_j = \beta_i, \qquad (i = 1, \ldots, n)$$

or in matrix notation

$$AX = b$$

where

$$A = (\alpha_{ij}), X^t = (x_1, \ldots, x_n), b^t = (\beta_1, \ldots, \beta_n)$$

If A is invertible, then

$$X = A^{-1}b = \frac{\text{adj } A}{\det A} b$$

or for $i = 1, 2, \ldots, n$

$$x_i = \frac{1}{\det A} \sum_{j=1}^{n} (-1)^{i+j} \det(A_{ji}) \beta_j$$

$$= \frac{\det A_i}{\det A}$$

60

where A_i is the matrix obtained from A by replacing its ith column with the column b, i.e.

$$A_i = \begin{pmatrix} \alpha_{11} & \cdots & \beta_1 & \cdots & \alpha_{1n} \\ \alpha_{21} & \cdots & \beta_2 & \cdots & \alpha_{2n} \\ \cdot & & \cdot & & \cdot \\ \cdot & & \cdot & & \cdot \\ \cdot & & \cdot & & \cdot \\ \alpha_{n1} & \cdots & \beta_n & \cdots & \alpha_{nn} \end{pmatrix}$$

with i marking the replaced column.

This is the well known **Cramer-solution** of linear equations.

EXAMPLE

$$\begin{aligned} x \quad\;\; - z &= 1 \\ 2x + y - z &= 1 \\ x + 2y + 5z &= 2 \end{aligned}$$

Using the inverse of the matrix A obtained in the above example we have

$$\begin{pmatrix} x \\ y \\ z \end{pmatrix} = \frac{1}{4} \begin{pmatrix} 7 & -2 & 1 \\ -11 & 6 & -1 \\ 3 & -2 & 1 \end{pmatrix} \begin{pmatrix} 1 \\ 1 \\ 2 \end{pmatrix} = \begin{pmatrix} \frac{7}{4} \\ -\frac{7}{4} \\ \frac{3}{4} \end{pmatrix}$$

or alternatively, Cramer's solution gives

$$x = \frac{1}{4} \det \begin{pmatrix} 1 & 0 & -1 \\ 1 & 1 & -1 \\ 2 & 2 & 5 \end{pmatrix} = \frac{7}{4} \quad y = \frac{1}{4} \det \begin{pmatrix} 1 & 1 & -1 \\ 2 & 1 & -1 \\ 1 & 2 & 5 \end{pmatrix} = -\frac{7}{4}$$

$$z = \frac{1}{4} \det \begin{pmatrix} 1 & 0 & 1 \\ 2 & 1 & 1 \\ 1 & 2 & 2 \end{pmatrix} = \frac{3}{4}$$

Exercises 2.4

1. Use the methods of this section to find the inverses of the following matrices

(i) $\begin{pmatrix} -2 & 3 & 2 \\ 6 & 0 & 3 \\ 4 & 1 & -1 \end{pmatrix}$ (ii) $\begin{pmatrix} 1 & -1 & 1 \\ 2 & 1 & 1 \\ 5 & -2 & 1 \end{pmatrix}$ (iii) $\begin{pmatrix} 1 & 2 & 3 \\ 1 & 3 & 5 \\ 1 & 5 & 12 \end{pmatrix}$

2. Use Cramer's method to solve the following systems of linear equations

(i)
$$x - y + 2z = 1$$
$$2x + y + z = 2$$
$$x - 3y + z = 1$$

(ii)
$$2x + 3y - 5z = 4$$
$$x + 7y - 2z = 1$$
$$5x - 11y + 2z = -2$$

3. For what values of t will the following matrix be non-invertible? For all other values of t, what is the inverse?

$$\begin{pmatrix} 1 & t & 0 \\ 0 & 1 & -1 \\ t & 0 & 1 \end{pmatrix}$$

4. If A is an $n \times n$ matrix, prove that

$$\det(\operatorname{adj} A) = (\det A)^{n-1}$$

5. Find the adjoint of A, where

$$A = \begin{pmatrix} x+1 & 0 & -1 \\ 0 & x+1 & -2 \\ 1 & 1 & x-2 \end{pmatrix}$$

If

$$B = \begin{pmatrix} 1 & 0 & 0 \\ 0 & 1 & 0 \\ 1 & 1 & x-1 \end{pmatrix}, \text{ for what complex values of } x \text{ is the}$$

product of adj A and B non-invertible?

CHAPTER 3

Vector Spaces

In this and the remaining chapters the following notation will be used:

> K is an arbitrary field
> **R** is the field of real numbers
> **C** is the field of complex numbers
> **Q** is the field of rational numbers.

3.1 Introduction

In physical connections, a vector is usually defined to be a "physical quantity which has magnitude and direction." As the work of this chapter involves a generalization of a vector, we first attempt to define more precisely a vector in the plane and in space.

We shall first consider the case of a plane. By a point in the plane we mean an ordered pair (a, b) of real numbers (i.e. each point is represented by coordinates a and b relative to a rectangular coordinate system). The origin O is the point $(0, 0)$. A **directed line segment** from the point P to the point Q, denoted by $\overrightarrow{PQ}$, is a line with initial point P and end point Q, that is, a directed line segment is an ordered pair of points (see Figure 1).

If $P = (p_1, p_2), Q = (q_1, q_2), R = (r_1, r_2), S = (s_1, s_2)$ are four points in the plane, the two directed line segments $\overrightarrow{PQ}$ and $\overrightarrow{RS}$ are said to be equivalent if

$$(q_1 - p_1, q_2 - p_2) = (s_1 - r_1, s_2 - r_2)$$

This relation is an equivalence relation on the set of all directed line segments in the plane, for it is clearly reflexive and symmetric and if also $\overrightarrow{RS}$ is equivalent to $\overrightarrow{TU}$, where $T = (t_1, t_2), U = (u_1, u_2)$ are two further points in the plane, then

$$(q_1 - p_1, q_2 - p_2) = (s_1 - r_1, s_2 - r_2) = (u_1 - t_1, u_2 - t_2)$$

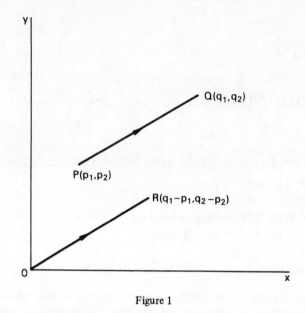

Figure 1

and the relation is also transitive. Now, it is easily seen that every directed line segment is equivalent to a unique directed line segment with initial point the origin O; i.e. if $P = (p_1, p_2)$, $Q = (q_1, q_2)$ then $\overrightarrow{PQ}$ is equivalent to $\overrightarrow{OR}$, where $R = (q_1 - p_1, q_2 - p_2)$. Thus, each equivalence class of directed line segments can be represented by a directed line segment with initial point the origin O. A **vector** $\overrightarrow{OP}$ in the plane is defined to be the directed line segment with initial point O and end point P. A vector is completely determined by the co-ordinates of its end point, thus a vector $\overrightarrow{OP}$ will be denoted by (p_1, p_2), the co-ordinates of the point P. Let $V_2(\mathbf{R})$ denote the set of all vectors in the plane, i.e.

$$V_2(\mathbf{R}) = \{(p_1, p_2) | p_1, p_2 \in \mathbf{R}\}$$

There is a well known rule, called the parallelogram rule, for adding vectors. If $\overrightarrow{OP}$ and $\overrightarrow{OQ}$ are vectors, a vector $\overrightarrow{OR}$ corresponds to these two vectors in the following way:

64

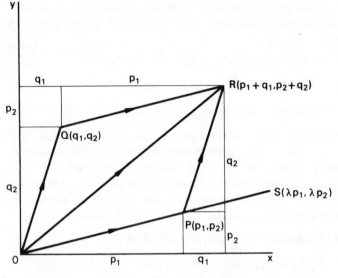

Figure 2

Let $\overrightarrow{QR}$ be the directed line segment with initial point Q which is equivalent to $\overrightarrow{OP}$ (i.e. QR is parallel to OP and with the same length and $\overrightarrow{PR}$ is the directed line segment equivalent to $\overrightarrow{OQ}$). Then $\overrightarrow{OR}$ is the sum of $\overrightarrow{OP}$ and $\overrightarrow{OQ}$, i.e.

$$\overrightarrow{OP} + \overrightarrow{OQ} = \overrightarrow{OR}$$

or

$$(p_1, p_2) + (q_1, q_2) = (p_1 + q_1, p_2 + q_2) \tag{1}$$

The multiplication of a vector $\overrightarrow{OP}$ by a real number λ can be dealt with similarly, $\lambda \overrightarrow{OP}$ is the vector $\overrightarrow{OS}$ of length $|\lambda| OP$ in the same direction as $\overrightarrow{OP}$ if λ is positive and in the opposite direction if λ is negative; this gives

$$\lambda (p_1, p_2) = (\lambda p_1, \lambda p_2) \tag{2}$$

Thus, on $V_2(\mathbf{R})$ addition is defined by the rule (1) and scalar multiplication by elements of $\mathbf{R}$ by (2).

Directed line segments and vectors in space (or 3-space) can be defined similarly. If we let $V_3(\mathbf{R})$ denote the set of all vectors in space, or

$$V_3(\mathbf{R}) = \{(\alpha, \beta, \gamma) | \alpha, \beta, \gamma \in \mathbf{R}\}$$

65

then similar geometric conditions will give the rule for addition in $V_3(\mathbf{R})$ and scalar multiplication of elements of $V_3(\mathbf{R})$ by scalars in $\mathbf{R}$, namely

$$(\alpha, \beta, \gamma) + (\alpha', \beta', \gamma') = (\alpha + \alpha', \beta + \beta', \gamma + \gamma')$$

and

$$\lambda(\alpha, \beta, \gamma) = (\lambda\alpha, \lambda\beta, \lambda\gamma)$$

With the pattern now developing, we can proceed to define $V_n(\mathbf{R})$ for $n \geqslant 4$, except that for these higher values of n we do not have a geometrical interpretation of our "vectors". We shall see in later sections of this chapter and book that the generalization proves to be a useful one. Indeed, rather than restricting ourselves to the real numbers only, we can work over any field K. So we give the following

DEFINITION 3.1 *If K is an arbitrary field, let*

$$V_n(K) = \{(\alpha_1, \alpha_2, \ldots, \alpha_n) \mid \alpha_i \in K \qquad (i = 1, \ldots, n)\}$$

*Define **addition** of elements in $V_n(K)$ by setting*

$$(\alpha_1, \alpha_2, \ldots, \alpha_n) + (\beta_1, \beta_2, \ldots, \beta_n) = (\alpha_1 + \beta_1, \alpha_2 + \beta_2, \ldots, \alpha_n + \beta_n)$$

*and **scalar multiplication** of elements of $V_n(K)$ by elements of K by*

$$\alpha(\alpha_1, \alpha_2, \ldots, \alpha_n) = (\alpha\alpha_1, \alpha\alpha_2, \ldots, \alpha\alpha_n)$$

for all $\alpha, \alpha_i, \beta_i \ (i = 1, \ldots, n) \in K$.

Having given these definitions, we can show that addition and scalar multiplication satisfy many rules which allow us to manipulate these vectors according to the usual laws of algebra. These are expressed in the following theorem.

THEOREM 3.2 *If $u = (\alpha_1, \ldots, \alpha_n), v = (\beta_1, \ldots, \beta_n), w = (\gamma_1, \ldots, \gamma_n)$ are arbitrary elements of $V_n(K)$ and α, β are arbitrary elements of K then*
  (i) *$u + v \in V_n(K)$*
  (ii) *$u + (v + w) = (u + v) + w$*
  (iii) *$u + v = v + u$*
  (iv) *there exists a $0 \in V_n(K)$ such that $u + 0 = u$ for all $u \in V_n(K)$*
  (v) *for each $u = (\alpha_1, \ldots, \alpha_n) \in V_n(K)$ there exists a*
      $-u = (-\alpha_1, \ldots, -\alpha_n) \in V_n(K)$ such that $u + (-u) = 0$
  (vi) *$\alpha u \in V_n(K)$*
  (vii) *$(\alpha + \beta)u = \alpha u + \beta u$*
  (viii) *$\alpha(u + v) = \alpha u + \alpha v$*

(ix) $\alpha(\beta u) = (\alpha\beta)u$

(x) $1u = u.$

PROOF   The proof is elementary; we shall prove some parts, the reader should verify the remainder, for example

$$
\begin{aligned}
u + (v + w) &= (\alpha_1, \ldots, \alpha_n) + \{(\beta_1, \ldots, \beta_n) + (\gamma_1, \ldots, \gamma_n)\} \\
&= (\alpha_1, \ldots, \alpha_n) + (\beta_1 + \gamma_1, \ldots, \beta_n + \gamma_n) \\
&= (\alpha_1 + (\beta_1 + \gamma_1), \ldots, \alpha_n + (\beta_n + \gamma_n)) \\
&= ((\alpha_1 + \beta_1) + \gamma_1, \ldots, (\alpha_n + \beta_n) + \gamma_n) \\
&= (u + v) + w
\end{aligned}
$$

which proves (ii). The $0 \in V_n(K)$ required in (iv) is the n-tuple

$0 = (0, 0, \ldots, 0)$ and in (v)

$$
\begin{aligned}
(\alpha_1, \ldots, \alpha_n) + (-\alpha_1, \ldots, -\alpha_n) &= (\alpha_1 + (-\alpha_1), \ldots, \alpha_n + (-\alpha_n)) \\
&= (0, \ldots, 0) = 0 \quad \blacksquare
\end{aligned}
$$

We shall not develop further our study of $V_n(K)$ at this stage. Indeed just as $V_2(\mathbf{R})$ and $V_3(\mathbf{R})$ were used to motivate the definition of $V_n(K)$ for arbitrary n and arbitrary fields K, we shall use (i)-(x) in Theorem 3.2 to motivate the definition of a general abstract structure, called a vector space of which $V_n(K)$ is merely an example, albeit a very important and typical example as will be seen before the end of this chapter.

3.2 Definition and Examples of Vector Spaces

DEFINITION 3.3   *A set V is called a **vector space over the field K** or a **K-space** if*

(a) (i) *V is closed with respect to a binary operation called addition* (+), *i.e. if u, $v \in V$, then $u + v \in V$,*

  (ii) *(commutative axiom) $u + v = v + u$ for all u, $v \in V$,*

  (iii) *(associative axiom) $u + (v + w) = (u + v) + w$ for all u, v, w \in V,*

  (iv) *there exists an element $0 \in V$, called the zero element, such that $u + 0 = 0 + u = u$ for all $u \in V$*

  (v) *for every $v \in V$, there exists an element $(-v) \in V$ such that $v + (-v) = 0 = (-v) + v,$*

(b) (i) *for every $\alpha \in K$, $v \in V$ an element αv called the **scalar multiple of v by α**, is defined and $\alpha\, v \in V$.*

(ii) $\alpha(u + v) = \alpha u + \alpha v$, *for all $\alpha \in K$, $u, v \in V$,*

(iii) $(\alpha + \beta)v = \alpha v + \beta v$, *for all $\alpha, \beta \in K$, $v \in V$,*

(iv) $\alpha(\beta v) = (\alpha\beta)v$, *for all $\alpha, \beta \in K$, $v \in V$,*

(v) $1v = v$, *for all $v \in V$.*

EXAMPLES

1. $V_n(K) = \{(\alpha_1, \ldots, \alpha_n) | \alpha_i \in K \,(i = 1, \ldots, n)\}$

was shown to be a K-space in the previous section.

2. Let $P(K)$ be the set of all polynomials in an indeterminate x with coefficients from K, i.e.

$$P(K) = \{\alpha_0 + \alpha_1 x + \ldots + \alpha_m x^m | \alpha_m \neq 0, \alpha_i \in K \,(i = 1, \ldots, m)\}$$

$+$ is the usual addition of polynomials and if $\alpha \in K$, then scalar multiplication by α is defined by

$$\alpha(\alpha_0 + \alpha_1 x + \ldots + \alpha_m x^m) = \alpha\alpha_0 + \alpha\alpha_1 x + \ldots + \alpha\alpha_m x^m$$

Then $P(K)$ is a K-space. To verify this all the axioms in definition 3.3 must be shown to be satisfied. In general, the verification is elementary, as in for example, 3.3 (a)(ii), if $f(x) = \alpha_0 + \alpha_1 x + \ldots + \alpha_m x^m$ and $g(x) = \beta_0 + \beta_1 x + \ldots + \beta_n x^n$, assuming, without loss of generality, that $m \geqslant n$, then $g(x) = \beta_0 + \beta_1 x + \ldots + \beta_n x^n + \ldots + \beta_m x^m$, where $\beta_{n+1} = \ldots = \beta_m = 0$. Then

$$\begin{aligned}
f(x) + g(x) &= (\alpha_0 + \beta_0) + (\alpha_1 + \beta_1)x + \ldots + (\alpha_m + \beta_m)x^m \\
&= (\beta_0 + \alpha_0) + (\beta_1 + \alpha_1)x + \ldots + (\beta_m + \alpha_m)x^m \\
&= g(x) + f(x)
\end{aligned}$$

3. Let $P_n(K)$ be the set of polynomials of degree $\leqslant n$ with coefficients in K, i.e.

$$P_n(K) = \{\alpha_0 + \alpha_1 x + \ldots + \alpha_n x^n | \alpha_i \in K \qquad (i = 1, \ldots, n)\}$$

If addition $(+)$ and scalar multiplication by elements of K are defined as in Example 2, an elementary verification again shows that $P_n(K)$ is a K-space.

4. Let X be an arbitrary set and K an arbitrary field. Let

$$K^X = \{f : X \to K\}$$

be the set of all mappings (functions) of X into K. If $f, g \in K^X$, $\alpha \in K$, then addition and scalar multiplications are defined on K^X by

$$(f+g)(x) = f(x) + g(x) \qquad \text{for all } x \in X$$

$$(\alpha f)(x) = \alpha f(x) \qquad \text{for all } x \in X$$

Then, it can be easily verified that K^X is a K-space.

Some important special cases are given in the next example.

5. (i) If $X = K = \mathbf{R}$, then $\mathbf{R}^{\mathbf{R}}$ is the $\mathbf{R}$-space of real valued functions defined on $\mathbf{R}$.

(ii) If X is the closed interval $[a,b] = \{x \in \mathbf{R} \mid a \leqslant x \leqslant b\}$ where a,b are any real numbers and $K = \mathbf{R}$, then $\mathbf{R}^{[a,b]}$ is the $\mathbf{R}$-space of real valued functions defined on the closed interval $[a,b]$.

(iii) If $X = \mathbf{N} = \{0,1,2 \ldots\}$ and $K = \mathbf{R}$ then $\mathbf{R}^{\mathbf{N}}$ is the $\mathbf{R}$-space of all sequences $\{\alpha_n\}$ (see Exercise 3.3, No. 6).

6. $\mathbf{C}$ is an $\mathbf{R}$-space since $\mathbf{C} = \{(\alpha, \beta) \mid \alpha, \beta \in \mathbf{R}\} = V_2(\mathbf{R})$.

7. Every field K is a K-space. Scalar multiplication of elements of K by elements of K is simply "ordinary" multiplication in K. Thus all the axioms for a K-space are covered by the defining axioms of a field. In particular $\mathbf{C}$ is a $\mathbf{C}$-space. Examples 6 and 7 show that the same set may be regarded as a vector space over more than one field.

8. Let $M_{m,n}(K)$ be the set of all $m \times n$ matrices over K. Addition of matrices and scalar multiplication of matrices by elements of K have been defined in Chapter I, §1.4, where most of the axioms for vector spaces were verified. The remaining axioms are easily verified and $M_{m,n}(K)$ is a K-space. $M_n(K)$ denotes the vector space of all $n \times n$ matrices over K.

The examples of vector spaces given above have come from various branches of mathematics, geometry, algebra, analysis and differential equations. The theory of vector spaces is therefore potentially of significance in the whole of mathematics. **The notation $V_n(K)$, $P(K)$, $P_n(K)$, R^R, $R^{[a,b]}$, $M_{m,n}(K)$, $M_n(K)$ for these vector spaces will be standard for the remainder of this book.**

We now prove some elementary consequences of the definition of a vector space.

THEOREM 3.4   *Let V be a K-space. Then*

(a) *the zero element* $0 \in V$ *defined in 3.3 (a) (iv) is unique*
(b) *if* $v \in V$, *then* $(-v)$ *defined in 3.3 (a)(v) is unique*
(c) *if* $u, v, w \in V$ *and* $u + v = u + w$ *then* $v = w$
(d) *if* $u, v \in V$ *then* $u + x = v$ *has a unique solution* $v - u$ *in* V
(e) $-(-v) = v$ *for all* $v \in V$
(f) $0.v = 0$ *for all* $v \in V$
(g) $-(\alpha v) = (-\alpha) v = \alpha(-v)$ *and* $(-\alpha)(-v) = \alpha v$ *for all* $\alpha \in K$, $v \in V$.

PROOF   (a) Let $0'$ also be a zero element, then

$$v + 0' = v = 0' + v \quad \text{for all} \quad v \in V$$

Thus, in particular

$$0 = 0 + 0' = 0' + 0 = 0'$$

since $v + 0 = v$ for all $v \in V$, i.e. $0 = 0'$

(b) Let v' also be an additive inverse of v, i.e.

$$v + v' = v' + v = 0$$

and $v + (-v) = (-v) + v = 0$

Then $v' = v' + 0 = v' + (v + (-v)) = (v' + v) + (-v)$

$$= 0 + (-v) = -v$$

(c) If $u + v = u + w$ then

$$(-u) + (u + v) = (-u) + (u + w)$$

i.e. $((-u) + u) + v = ((-u) + u) + w$

i.e. $0 + v = 0 + w$ i.e. $v = w$

(d) Since $u + ((-u) + v) = (u + (-u)) + v = 0 + v = v$, $u + x = v$ has at least one solution. This solution is unique, for if x and x' are solutions then $u + x = u + x'$ and by (c) $x = x'$. This unique solution is denoted by $v - u$.

(e) If $v \in V$, we have $(-v) + (-(-v)) = 0 = (-v) + v$, which by (c) implies $-(-v) = v$.

(f) If $v \in V$, then $0.v = (0 + 0).v = 0.v + 0.v$, but $0.v = 0 + 0.v$ and by (a) $0 = 0.v$.

(g) If $\alpha \in K$, $v \in V$ then $(\alpha v) + (-(\alpha v)) = 0$ and $v + (-(v)) = 0$ implies $\alpha v + \alpha(-v) = \alpha.0 = 0$ and by (b) $\alpha(-v) = -(\alpha v)$. Also, $\alpha + (-\alpha) = 0$ and so $(\alpha + (-\alpha)).v = 0.v = 0$ by (f). Thus $\alpha v + (-\alpha)v = 0 = \alpha v + (-(\alpha v))$ and so $(-\alpha)(v) = -(\alpha v)$. Finally, $(-\alpha)(-v) = \alpha(-(-v)) = \alpha v$ by (e).   ∎

70

Note   The reader should determine exactly which of the defining axioms for a vector space are used in the various parts of the above proof.

3.3 Subspaces

Let V be a K-space.

DEFINITION 3.5   *A non-empty subset W of V is called a **subspace** of V if W is itself a K-space with the same definition of addition and scalar multiplication as in V.*

By inspecting the defining axioms for a vector space, we note that some of the axioms are automatically true in W if W is a subset of V, i.e. 3.3 (a)(ii), (iii) and (b)(ii), (iii), (iv) and (v). Therefore, to verify that a subset W of V is a subspace we need only show that

(i)   if $u, v \in W$ then $u + v \in W$
(ii) $0 \in W$
(iii)   for each $v \in W$, $-v \in W$
(iv)   if $\alpha \in K$, $v \in W$ then $\alpha v \in W$.

In fact, we can prove

LEMMA 3.6   *A non-empty subset W of V is a subspace of V if and only if $\alpha u + v \in W$ for all $\alpha \in K$, $u, v \in W$.*

PROOF   If W is a subspace of V, then by definition, if $\alpha \in K$, $u, v \in W$, then $\alpha u \in W$ and $\alpha u + v \in W$. Conversely, since W is non-empty, there exists an element $u \in W$ and thus $(-1)u + u = -u + u = 0 \in W$ and hence if $\alpha \in K$, $v \in V$ $\alpha . v + 0 = \alpha v \in W$ and in particular $-1 . v = -v \in W$. Finally, if $u, v \in W$, $1u + v = u + v \in W$. That is (i)-(iv) above have been verified, and W is a subspace of V. ∎

EXAMPLES

1. Every vector space V has two subspaces, namely V and $\{0\}$. Every other subspace is called a **proper** subspace.

2. Let $W = \{(0, \alpha_2, \ldots, \alpha_n) \mid \alpha_i \in K \ (i = 2, \ldots, n)\} \subset V_n(K)$, then W is a subspace of $V_n(K)$, since

(i) W is non-empty, for example $(0, \ldots, 0) \in W$
(ii)   if $\alpha \in K$, $u = (0, \alpha_2, \ldots, \alpha_n)$, $v = (0, \beta_2, \ldots, \beta_n) \in W$, then $\alpha u + v = (0, \alpha\alpha_2 + \beta_2, \ldots, \alpha\alpha_n + \beta_n) \in W$.

Let $W = \{(1 + \alpha_2, \alpha_2, \ldots, \alpha_n) \mid \alpha_i \in K \ (i = 2, \ldots, n)\} \subseteq V_n(K)$, since, for example, $(2, 1, 0, \ldots, 0) \in W$ but $2(2, 1, 0, \ldots, 0) = (4, 2, 0, \ldots, 0) \notin W$, W is **not** a subspace of $V_n(K)$.

3. Let $M_n^{(s)}(K) = \{A \in M_n(K) \mid A^t = A\} \subset M_n(K)$, then $M_n^{(s)}(K)$ is a subspace of $M_n(K)$ since $0 \in M_n^{(s)}(K)$ and if $\alpha \in K, A, B \in M_n^{(s)}(K)$ then

$$(\alpha A + B)^t = \alpha A^t + B^t = \alpha A + B$$

i.e. $\alpha A + B \in M_n^{(s)}(K)$.

4. Let V be the set of solutions of the system of m linear homogeneous equations in n variables $\sum\limits_{j=1}^{n} \alpha_{ij} x_j = 0 \, (i = 1, \ldots, m)$, or $AX = 0$, where $A = (\alpha_{ij}), X = (x_1, \ldots, x_n)^t$. Then V is a subspace of $V_n(K)$, called the **solution space** of the system, i.e. $0 \in V$ and if $\alpha \in K, X, Y \in V$ then

$$A(\alpha X + Y) = \alpha AX + AY$$

$$= 0$$

and $\alpha X + Y \in V$. This solution space will be considered later when the vector space theory to be developed will be applied to linear equations. Indeed, the problem of determining the solution of a system of linear homogeneous equations may be rephrased as the determination of the solution space of the system.

5. Let $C[a,b]$ be the set of real valued continuous functions defined on the closed interval $[a,b]$, then $C[a,b]$ is a subspace of $\mathbf{R}^{[a,b]}$. In fact, if $f,g \in C[a,b]$, $\alpha \in \mathbf{R}$, then $\alpha f + g \in C[a,b]$. This follows from theorems in elementary calculus, which prove that if f and g are continuous functions then $\alpha f + g$ are continuous (see for example A.S.-T. Lue, *Basic Pure Mathematics II*, VNR New Mathematics Library 5, p.51).

Let $C'[a,b]$ be the set of all continuously differential functions defined on a closed interval $[a,b]$, i.e. the set of all functions which possess a continuous derivative on every point in $[a,b]$, then again theorems in elementary calculus show that $C'[a,b]$ is a subspace of $\mathbf{R}^{[a,b]}$. The question arises whether $C'[a,b]$ and $C[a,b]$ are the same spaces; in fact, $C'[a,b] \subset C[a,b]$, as a differentiable function is continuous. The inclusion is strict; since, for example, the function $f(x) = |x|$ is continuous but has no derivative at $x = 0$. Sometimes, we are required to deal with functions which may be differentiated a number of times; we therefore denote by $C^n[a,b]$ the set of all n times continuously differentiable functions on $[a,b]$ and by $C^\infty[a,b]$ the set which may be differentiated any number of times. Again, it may be verified that $C^n[a,b]$ $(n \geqslant 1)$ and $C^\infty[a,b]$ are $\mathbf{R}$-spaces (i.e. they are subspaces of $\mathbf{R}^{[a,b]}$) and that if $m \geqslant n$, $C^m[a,b]$ is a subspace of $C^n[a,b]$. We shall add $C[a,b]$, $C^n[a,b]$ and $C^\infty[a,b]$ to the list of

standard notation for vector spaces given on p. 69. In addition
$C(\mathbf{R}), C^n(\mathbf{R}), C^\infty(\mathbf{R})$ will be used when $[a,b] = \mathbf{R}$.

6. Let V be the subset of $C^2[a,b]$ which are solutions of the fixed
homogeneous linear differential equations

$$\frac{d^2 y}{dx^2} + a(x)\frac{dy}{dx} + b(x)y = 0,$$

where $a(x), b(x) \in C[a,b]$. If $f, g \in V$, $\alpha \in \mathbf{R}$ then

$$\frac{d^2}{dx^2}(f{+}g)(x) + a(x)\frac{d}{dx}(f{+}g)(x) + b(x)(f{+}g)(x)$$

$$= \left(\frac{d^2}{dx^2}f(x) + a(x)\frac{d}{dx}f(x) + b(x)f(x)\right)$$

$$+\left(\frac{d^2}{dx^2}g(x) + a(x)\frac{d}{dx}g(x) + b(x)g(x)\right)$$

$$= 0$$

where again some theorems in elementary calculus have been used (*loc.
cit* p.57). Thus $f + g \in V$ and similarly $\alpha f \in V$ and V is a subspace of
$C^2[a,b]$. V is a *solution space* of the above differential equations, which
is an important example of vector spaces. Sometimes additional
"boundary conditions" must be satisfied, for example, if $U \leqslant C^2[0,\pi]$
is the solution space of the differential equation

$$\frac{d^2 y}{dx^2} + y = 0$$

with boundary conditions $y(0) = 0, y(\pi) = 0$. Then it is easily verified
that U is a subspace of $C^2[0,\pi]$.

We now consider ways of constructing some important subspaces of
a vector space. If U and W are subspaces of V, then the intersection of
U and W, $U \cap W$ is also subspace, for $0 \in U \cap W$ and if $\alpha \in K$,
$u, v \in U \cap W$ then $u, v \in U$ and $u, v \in W$ and $\alpha u + v \in U$ and
$\alpha u + v \in W$, which implies that $\alpha u + v \in U \cap W$. But, in general the
union of U and W, $U \cup W$ is not a subspace of V, for example,
$U = \{(\alpha, 0) \mid \alpha \in K\}$ and $W = \{(0, \beta) \mid \beta \in K\}$ are both subspaces of
$V_2(K)$, but for example $(1, 0) + (0, 1) = (1, 1) \notin U \cup W$. The position
is rescued by the definition of the **sum** of two subspaces, i.e.
$U + W = \{u + w \mid u \in U, w \in W\}$, which is easily shown to be a subspace

of V. In fact, this can be extended to cover the sum of more than two subspaces.

DEFINITION 3.7 *Let* $W_1, W_2, \ldots, W_k$ *be subspaces of a vector space* V *over* K. *Define*

$$W = W_1 + \ldots + W_k = \{w_1 + w_2 + \ldots + w_k | w_i \in W_i \\ (i = 1, \ldots, k)\}$$

then W *is called the **sum** of the subspaces* $W_1, \ldots, W_k$.

Then we can prove by using Lemma 3.6,

LEMMA 3.8 (i) $W_1 \cap W_2 \cap \ldots \cap W_k$ *is a subspace of* V

(ii) $W_1 + W_2 + \ldots + W_k$ *is a subspace of* V.

PROOF (i) Since $0 \in W_i (i = 1, \ldots, k), 0 \in W_1 \cap W_2 \cap \ldots \cap W_k$ and so it is non-empty. If $\alpha \in K, u, v \in W_1 \cap W_2 \cap \ldots \cap W_k$, then $u, v \in W_i (i = 1, \ldots, k)$ and $\alpha u + v \in W_i (i = 1, \ldots, k)$ and $\alpha u + v \in W_1 \cap W_2 \cap \ldots \cap W_k$.

(ii) $0 = 0 + \ldots + 0 \in W_1 + W_2 + \ldots + W_k$ since $0 \in W_i (i = 1, \ldots, k)$. Further, if $\alpha \in K, u, v \in W_1 + W_2 + \ldots + W_k$, then $u = u_1 + \ldots + u_k$, $v = v_1 + \ldots + v_k$, where $u_i, v_i \in W_i (i = 1, \ldots, k)$. Thus, $\alpha u_i + v_i \in W_i (i = 1, \ldots, k)$ and it follows that

$$\alpha u + v = \alpha(u_1 + \ldots + u_k) + (v_1 + \ldots + v_k)$$

$$= (\alpha u_1 + v_1) + \ldots + (\alpha u_k + v_k),$$

that is, $\alpha u + v \in W_1 + \ldots + W_k$. ∎

Now, let $S = \{v_1, \ldots, v_k\}$ be a finite subset of a vector space V. Let $U = \left\{ \sum_{i=1}^{k} \alpha_i v_i | \alpha_i \in K \right\}$. Then, we can prove

LEMMA 3.9 U *is a subspace of* V.

PROOF $0 \in U$, since $0 = 0 . v_1 + \ldots + 0 . v_k \in U$.

If $\alpha \in K, u, v \in U$ then $u = \sum_{i=1}^{k} \alpha_i v_i$ and $v = \sum_{i=1}^{k} \beta_i v_i$ for some $\alpha_i, \beta_i \in K (i = 1, \ldots, k)$. Then

$$\alpha u + v = \alpha \left(\sum_{i=1}^{k} \alpha_i v_i \right) + \left(\sum_{i=1}^{k} \beta_i v_i \right)$$

$$= \sum_{i=1}^{k} (\alpha \alpha_i + \beta_i) v_i \in U$$

and so, by Lemma 3.6, U is a subspace of V. ∎

74

DEFINITION 3.10   *If* $S = \{v_1, \ldots, v_k\} \subseteq V$, *then* $\sum\limits_{i=1}^{k} \alpha_i v_i$, *where*

$\alpha_i \in K$ *is called a **K-linear combination** of S. The subspace U constructed above is called the subspace **generated** (or **spanned**) by the subset S. We shall denote this by* $U = \langle S \rangle$ *or* $\langle v_1, \ldots, v_k \rangle$. *A subspace U of V which is generated by a finite subset of V is called a **finitely generated** subspace of V.* These ideas will be illustrated by examples to be considered in the next section.

(Definition 3.10 can be extended to cover infinite subsets. If S is a subset of V, let $U = \left\{ \sum\limits_{s \in S} \alpha_s s \,|\, \alpha_s \in K \right\}$, where in each sum, all but a finite number of the α_s are equal to zero. Then U is a subspace of V and is the subspace **generated** by S).

Exercises 3.3

1. Which of the following subsets of $V_n(\mathbf{R})$ $(n \geqslant 3)$ are vector spaces?
   (i) $\{(\alpha_1, \alpha_2, \ldots, \alpha_n) \,|\, \alpha_1 + \alpha_2 + \alpha_3 = 0\}$
   (ii) $\{(\alpha_1, \alpha_2, \ldots, \alpha_n) \,|\, \alpha_1 + \alpha_2 + \alpha_3 = 1\}$
   (iii) $\{(\alpha_1, \alpha_2, \ldots, \alpha_n) \,|\, \alpha_3 = \alpha_1 \alpha_2\}$
   (iv) $\{(\alpha_1, \alpha_2, \ldots, \alpha_n) \,|\, |\alpha_1| > 0\}$
   (v) $\{(\alpha_1, \alpha_2, \ldots, \alpha_n) \,|\, \alpha_1 - \alpha_2 = \alpha_2 - \alpha_3\}$

2. Which of the following subsets of $\mathbf{R}^{\mathbf{R}}$ are $\mathbf{R}$-spaces?
   (i) The set of all real polynomials of degree exactly n
   (ii) The set of all real polynomials of degree $< n$
   (iii) The set of all odd functions (i.e. $f(-x) = -f(x)$ for all $x \in \mathbf{R}$)
   (iv) The set of all differentiable functions.

3. Which of the following subsets of $C[0, 1]$ are $\mathbf{R}$-spaces?
   (i) $\{f \in C[0,1] \,|\, f(1) = 0\}$
   (ii) $\{f \in C[0,1] \,|\, f(1) = 1\}$
   (iii) $\{f \in C[0,1] \,|\, f(0) = f(1)\}$
   (iv) $\{f \in C[0,1] \,|\, \int_0^1 f(t)\, dt = 0\}$
   (v) $\{f \in C[0,1] \,|\, \int_0^1 f(t)\, dt = 1\}$

4. Which of the following subsets are subspaces of $M_2(\mathbf{R})$?
   (i) $\left\{ \begin{pmatrix} a & b \\ c & d \end{pmatrix} \,\middle|\, a = b \right\}$
   (ii) $\left\{ \begin{pmatrix} a & b \\ c & d \end{pmatrix} \,\middle|\, a + b = 1 \right\}$
   (iii) $\left\{ \begin{pmatrix} a & b \\ c & d \end{pmatrix} \,\middle|\, a = c = d \right\}$

(iv) $\{A \in M_2(\mathbf{R}) \mid \det A = 0\}$

(v) $\{A \in M_2(\mathbf{R}) \mid \det A = 1\}$

(vi) $\{A \in M_2(\mathbf{R}) \mid AB = BA \text{ for a fixed } B \in M_2(\mathbf{R})\}$

(vii) $\{A \in M_2(\mathbf{R}) \mid A^2 = A\}$.

5. Which of the following subsets are subspaces of $M_3(\mathbf{R})$

(i) $\{A \in M_3(\mathbf{R}) \mid \det A = 0\}$

(ii) $\left\{A \in M_3(\mathbf{R}) \mid \sum_{i=1}^{3} a_{ii} = 0\right\}$

(iii) $\{A \in M_3(\mathbf{R}) \mid A^t = -A\}$.

6. Let V be the set of all sequences $\{\alpha_n\} = (\alpha_0, \alpha_1, \ldots, \alpha_n, \ldots)$ of elements of $\mathbf{R}$. Define addition of sequences by

$$\{\alpha_n\} + \{\beta_n\} = \{\alpha_n + \beta_n\}$$

and scalar multiplication by

$$\alpha \{\alpha_n\} = \{\alpha \alpha_n\}$$

Prove that V is a vector space over $\mathbf{R}$.

7. Prove that the following are vector spaces over $\mathbf{R}$

(i) $\{\{\alpha_n\} \mid \alpha_{n+1} - \alpha_n = \alpha_{n+2} - \alpha_{n+1}, n \geqslant 0, \alpha_n \in \mathbf{R}\}$.

i.e. the set of all arithmetical progressions.

(ii) $\{\{\alpha_n\} \mid \alpha_{n+2} = \alpha_{n+1} + \alpha_n, n \geqslant 0, \alpha_n \in \mathbf{R}\}$,

i.e. the set of all Fibonacci sequences.

(iii) the set of all convergent sequences

i.e. $\{\{\alpha_n\} \mid$ for each $\epsilon > 0$ there exists an integer N such that $|\alpha_n - \alpha| < \epsilon$ for $n > N$ for some $\alpha\}$.

(iv) the set of all Cauchy sequences

$\{\{\alpha_n\} \mid$ for each $\epsilon > 0$ there exists an integer N such that $|\alpha_m - \alpha_n| < \epsilon$ for $m, n > N\}$.

3.4 Linear Independence, Basis and Dimension

Let V be a K-space.

DEFINITION 3.11   A subset $\{v_1, \ldots, v_k\}$ in V is **linearly dependent over K** if there exist $\alpha_1, \ldots, \alpha_k \in K$ (not all zero) such that

$$\alpha_1 v_1 + \alpha_2 v_2 + \ldots + \alpha_k v_k = 0$$

If not, $\{v_1, \ldots, v_k\}$ is **linearly independent over K**, or in other words, if $\alpha_1 v_1 + \alpha_2 v_2 + \ldots + \alpha_k v_k = 0$ implies that $\alpha_i = 0$ $(i = 1, \ldots, k)$, then $\{v_1, \ldots, v_k\}$ is linearly independent over K.

EXAMPLES

1. $\{(1, 0, 1)\ \ (1, -1, 1)\ \ (2, -1, 2)\ \ (0, 0, 1)\}$
is linearly dependent over **R** since

$$(1, 0, 1) + (1, -1, 1) - (2, -1, 2) + 0\,(0, 0, 1) = (0, 0, 0)$$

but $\{(1, 0, 1), (0, 0, 1)\} \subseteq V_3(\mathbf{R})$ is linearly independent over **R**,

for if $\alpha(1, 0, 1) + \beta(0, 0, 1) = (0, 0, 0)$,

then $\alpha = 0, \alpha + \beta = 0$, or $\alpha = 0, \beta = 0$

2. We have seen in examples 5 and 6 in §3.2 that **C** is an **R**-space and a
C-space.
$1 + i$ and i are linearly dependent over **C**, i.e. $1(1 + i) + (-1 + i)i = 0$,
but linearly independent over **R**, for if $\alpha(1 + i) + \beta i = 0$, $\alpha, \beta \in \mathbf{R}$, then
$\alpha = 0$ and $\alpha + \beta = 0$ or $\alpha = \beta = 0$.
(Definition 3.11 can also be extended to cover infinite subsets, a subset
$S \subseteq V$ is linearly independent over K if every finite subset of S is
linearly independent over K, otherwise S is linearly dependent over K).

DEFINITION 3.12 *A subset $S \subseteq V$ is a K-**basis** for V if*
   (i) *S generates V*
   (ii) *S is linearly independent over K.*

If $S = \{v_1, \ldots, v_k\}$ is a finite set, this means that if $v \in V$ then

$v = \sum\limits_{i=1}^{k} \alpha_i v_i$, where $\alpha_i \in K$ $(i = 1, \ldots, k)$ and if $\sum\limits_{i=1}^{k} \beta_i v_i = 0$,

$\beta_i \in K$ $(i = 1, \ldots, k)$ then $\beta_i = 0$ $(i = 1, \ldots, k)$.
We shall concentrate on this case, that is, finitely generated K-spaces.
Before illustrating these ideas with examples, we give a useful
determinantal criterion for the n columns of a square matrix to be
linearly independent. We prove

THEOREM 3.13 *If $A = (\alpha_{ij}) \in M_n(K)$ and $c_j = (\alpha_{1j}, \alpha_{2j}, \ldots, \alpha_{nj})$
$(j = 1, 2, \ldots, n)$ are the n columns of A then $\{c_1, c_2, \ldots, c_n\}$ is
linearly dependent over K if and only if $\det A = 0$.*

PROOF If $\{c_1, c_2, \ldots, c_n\}$ is linearly dependent over K, then there
exist $\alpha_1, \alpha_2, \ldots, \alpha_n \in K$ (not all zero) such that

$$\alpha_1 c_1 + \alpha_2 c_2 + \ldots + \alpha_n c_n = 0$$

Without loss of generality we assume that $\alpha_1 \neq 0$ then

$c_1 + \dfrac{\alpha_2}{\alpha_1} c_2 + \ldots + \dfrac{\alpha_n}{\alpha_1} c_n = 0$. Now, subtracting $\dfrac{\alpha_2}{\alpha_1}$ times the second column $+ \ldots + \frac{\alpha_n}{\alpha_1}$ times the nth column from the first column gives a matrix with first column zero. Thus by Remark 2 in §2.2, det $A = 0$.

If det $A = 0$, then by Corollary 3 to theorem 2.7, the system of linear equations

$$\sum_{j=1}^{n} \alpha_{ij} x_j = 0 \qquad (i = 1, 2, \ldots, n)$$

has a non-trivial solution $(x_1, x_2, \ldots, x_n)$, that is, there exist $x_1, x_2, \ldots, x_n \in K$ (not all zero) such that

$$x_1 c_1 + x_2 c_2 + \ldots + x_n c_n = 0$$

which shows that $\{c_1, c_2, \ldots, c_n\}$ is linearly dependent over K. ∎
By considering the transpose A^t of A, we obtain

COROLLARY *The n rows of a matrix $A \in M_n(K)$ are linearly dependent over K if and only if det $A = 0$.*

EXAMPLES

1. Let $e_1 = (1, 0, 0)$, $e_2 = (0, 1, 0)$, $e_3 = (0, 0, 1)$ then $\{e_1, e_2, e_3\}$ is an **R**-basis for the **R**-space $V_3(\mathbf{R})$, since
   (i) if $v = (\alpha_1, \alpha_2, \alpha_3)$ is an arbitrary element in $V_3(\mathbf{R})$, then

$$(\alpha_1, \alpha_2, \alpha_3) = \alpha_1(1, 0, 0) + \alpha_2(0, 1, 0) + \alpha_3(0, 0, 1)$$

that is $\{e_1, e_2, e_3\}$ generates $V_3(\mathbf{R})$.
   (ii) if $\alpha_1 e_1 + \alpha_2 e_2 + \alpha_3 e_3 = 0$, then $(\alpha_1, \alpha_2, \alpha_3) = (0, 0, 0)$, or $\alpha_1 = \alpha_2 = \alpha_3 = 0$, that is $\{e_1, e_2, e_3\}$ is linearly independent over **R**.

Let $v_1 = (1, 0, 1)$, $v_2 = (2, -1, 1)$, $v_3 = (4, 1, 1)$, then $\{v_1, v_2, v_3\}$ is also an **R**-basis for $V_3(\mathbf{R})$. It can easily be verified that

$$(\alpha_1, \alpha_2, \alpha_3) = \tfrac{1}{4} \{ (-2\alpha_1 + 2\alpha_2 + 6\alpha_3) v_1 + (\alpha_1 - 3\alpha_2 - \alpha_3) v_2 \\ + (\alpha_1 + \alpha_2 - \alpha_3) v_3 \}$$

that is, $\{v_1, v_2, v_3\}$ generates $V_3(\mathbf{R})$. Also, $\{v_1, v_2, v_3\}$ is linearly independent over **R**, for if

$$\alpha_1 v_1 + \alpha_2 v_2 + \alpha_3 v_3 = 0$$

where $\alpha_1, \alpha_2, \alpha_3 \in \mathbf{R}$, then

$$\alpha_1(1, 0, 1) + \alpha_2(2, -1, 1) + \alpha_3(4, 1, 1) = (0, 0, 0)$$

or $\qquad \alpha_1 + 2\alpha_2 + 4\alpha_3 = 0$

$$-\alpha_2 + \alpha_3 = 0$$

$$\alpha_1 + \alpha_2 + \alpha_3 = 0$$

Since $\det \begin{pmatrix} 1 & 2 & 4 \\ 0 & -1 & 1 \\ 1 & 1 & 1 \end{pmatrix} = 4 \neq 0$, the trivial solution $\alpha_1 = \alpha_2 = \alpha_3 = 0$

is the unique solution of this system (see Theorem 3.13).

This example shows that an **R**-basis for a vector space is not unique.

2. Let $e_1 = (1, 0, \ldots, 0)$, $e_2 = (0, 1, 0, \ldots, 0)$, \ldots,
$e_n = (0, 0, \ldots, 1) \in V_n(K)$. If $(\alpha_1, \alpha_2, \ldots, \alpha_n)$ is an arbitrary element
in $V_n(K)$, then

$$(\alpha_1, \ldots, \alpha_n) = \sum_{i=1}^{n} \alpha_i e_i$$

that is

$$\{e_1, \ldots, e_n\} \text{ generates } V_n(K)$$

If $\sum_{i=1}^{n} \alpha_i e_i = 0$ then $(\alpha_1, \ldots, \alpha_n) = (0, \ldots, 0)$, or
$\alpha_i = 0 \, (i = 1, \ldots, n)$, that is, $\{e_1, \ldots, e_n\}$ is linearly independent over
K. Thus $\{e_1, \ldots, e_n\}$ is a K-basis for $V_n(K)$, called the **standard K-basis**
for $V_n(K)$.

3. Let $V = M_2(\mathbf{R})$ and let

$$e_{11} = \begin{pmatrix} 1 & 0 \\ 0 & 0 \end{pmatrix}, \quad e_{12} = \begin{pmatrix} 0 & 1 \\ 0 & 0 \end{pmatrix},$$

$$e_{21} = \begin{pmatrix} 0 & 0 \\ 1 & 0 \end{pmatrix}, \quad e_{22} = \begin{pmatrix} 0 & 0 \\ 0 & 1 \end{pmatrix},$$

then $\{e_{11}, e_{12}, e_{21}, e_{22}\}$ is an **R**-basis for $M_2(\mathbf{R})$, since

(i) if $\begin{pmatrix} \alpha & \beta \\ \gamma & \delta \end{pmatrix} \in M_2(\mathbf{R})$ then

$$\begin{pmatrix} \alpha & \beta \\ \gamma & \delta \end{pmatrix} = \alpha e_{11} + \beta e_{12} + \gamma e_{21} + \delta e_{22}$$

and $\{e_{11}, e_{12}, e_{21}, e_{22}\}$ generates $M_2(\mathbf{R})$ over **R**.

(ii) to show that this set is linearly independent over **R**, consider

$$\alpha e_{11} + \beta e_{12} + \gamma e_{21} + \delta e_{22} = \begin{pmatrix} 0 & 0 \\ 0 & 0 \end{pmatrix}$$

or
$$\begin{pmatrix} \alpha & \beta \\ \gamma & \delta \end{pmatrix} = \begin{pmatrix} 0 & 0 \\ 0 & 0 \end{pmatrix}$$

which implies that $\alpha = \beta = \gamma = \delta = 0$.

4. We now show similarly that $S = \{e_{ij} \mid i, j = 1, \ldots, n\}$ is a K-basis for $M_n(K)$, where e_{ij} is the $n \times n$ matrix with 1 in the (i, j)-position and zero elsewhere. If (α_{ij}) is an arbitrary matrix in $M_n(K)$ then

$$(\alpha_{ij}) = \sum_{i, j = 1}^{n} \alpha_{ij} e_{ij}$$

and so S generates $M_n(K)$ over K. Consider

$$\sum_{i, j = 1}^{n} \alpha_{ij} e_{ij} = 0$$

where $\alpha_{ij} \in K$ $(i, j = 1, \ldots, n)$, then $(\alpha_{ij}) = 0$, which implies that $\alpha_{ij} = 0$ $(i, j = 1, \ldots, n)$ and S is linearly independent over K.

5. Let $V = V_2(\mathbf{C})$. In Example 2, we saw that $\{(1, 0), (0, 1)\}$ is a $\mathbf{C}$-basis for V. We now show that $S = \{(1, 0), (i, 0), (0, 1), (0, i)\}$ is an $\mathbf{R}$-basis for $V_2(\mathbf{C})$ regarded as an $\mathbf{R}$-space.
$V_2(\mathbf{C}) = \{(\alpha_1 + \alpha_2 i, \beta_1 + \beta_2 i) \mid \alpha_1, \alpha_2, \beta_1, \beta_2 \in \mathbf{R}\}$. Since
$(\alpha_1 + \alpha_2 i, \beta_1 + \beta_2 i) = \alpha_1(1, 0) + \alpha_2(i, 0) + \beta_1(0, 1) + \beta_2(0, i)$, for all
$\alpha_1, \alpha_2, \beta_1, \beta_2 \in \mathbf{R}$, we have that S generates $V_2(\mathbf{C})$ over $\mathbf{R}$. Consider

$$\alpha_1(1, 0) + \alpha_2(i, 0) + \beta_1(0, 1) + \beta_2(0, i) = (0, 0)$$

where $\alpha_1, \alpha_2, \beta_1, \beta_2 \in \mathbf{R}$, then

which implies that $\alpha_1 + \alpha_2 i = 0$, $\beta_1 + \beta_2 i = 0$ or $\alpha_1 = \alpha_2 = \beta_1 = \beta_2 = 0$
and S is also linearly independent over $\mathbf{R}$.

6. It can be proved in a similar manner that
$\{(1, 0, 0, \ldots, 0), (0, 1, 0, \ldots, 0), \ldots, (0, 0, \ldots, 0, 1), (i, 0, \ldots, 0),$
$(0, i, 0, \ldots, 0), \ldots (0, 0, \ldots, 0, i)\}$ is an $\mathbf{R}$-basis for $V_n(\mathbf{C})$ regarded as an $\mathbf{R}$-space.

7. Find an $\mathbf{R}$-basis for the solution space of the system of linear equations
$$x - 2y + 3z - w = 0$$
$$2x + y - z + w = 0$$

The reduced echelon matrix of this system is easily shown to be

$$\begin{pmatrix} 1 & 0 & \frac{1}{5} & \frac{1}{5} \\ 0 & 1 & -\frac{7}{5} & \frac{3}{5} \end{pmatrix}$$

and the equations become

$$x = -\tfrac{1}{5}z - \tfrac{1}{5}w$$
$$y = \tfrac{7}{5}z - \tfrac{3}{5}w$$

If we give z, w the arbitrary values λ, μ respectively, then the general solution of the system is

$$(x, y, z, w) = (-\lambda - \mu,\ 7\lambda - 3\mu,\ 5\lambda,\ 5\mu)$$
$$= \lambda\,(-1, 7, 5, 0) + \mu\,(-1, -3, 0, 5)$$

i.e. the solution space is $<(-1, 7, 5, 0), (-1, -3, 0, 5)>$ which is clearly an **R**-basis.

8. Find an **R**-basis for the solution space of the differential equation

$$\frac{\mathrm{d}^2 y}{\mathrm{d}x^2} + y = 0 \quad \text{or} \quad y'' + y = 0 \tag{*}$$

where $y \in C^\infty(\mathbf{R})$. This may be solved using elementary calculus as follows: Multiplying throughout by y' gives

$$y''y' + yy' = 0$$

and integrating

$$y'^2 + y^2 = c^2$$

where c is an arbitrary constant. Thus we have that

$$y' = \sqrt{c^2 - y^2} \quad \text{or} \quad \frac{\mathrm{d}y}{\sqrt{c^2 - y^2}} = \mathrm{d}x$$

which may be integrated to give

$$\sin^{-1} \frac{y}{c} = x + d$$

or

$$y = c \sin(x + d)$$

where d is also an arbitrary constant. Thus, we have that

$$y = c \cos d \sin x + c \sin d \cos x$$
$$= \gamma \sin x + \delta \cos x$$

where $\gamma, \delta \in \mathbf{R}$ and the solution space of (*) is $<\sin x, \cos x>$.

81

Now $\{\sin x, \cos x\}$ is linearly independent, for if

$$\alpha \sin x + \beta \cos x = 0$$

with $\alpha, \beta \in \mathbf{R}$, then by putting $x = 0$ and $x = \pi/2$, we obtain that $\beta = 0$ and $\alpha = 0$ respectively. Thus $\{\sin x, \cos x\}$ is an $\mathbf{R}$-basis for the solution space of (*).

Remark Note the similarity in approach to the problems of solutions of linear homogeneous equations and linear differential equations. Whereas we have given a systematic method for obtaining a K-basis for the solution space of a system of linear homogeneous equations the corresponding problem is not considered for differential equations. The first principle method for solving Example 8 is given for illustrative reasons only — for a comprehensive treatment of the application of linear algebra to differential equations and analysis see *An Introduction to Linear Analysis* by D.L. Kneider, R.G. Kuller, D.R. Ostberg and F.W. Perkins (Addison-Wesley).

Let $\mathscr{B} = \{v_1, v_2, \ldots, v_n\}$ be a K-basis for V. If $v \in V$, then
$$v = \sum_{i=1}^{n} \alpha_i v_i,$$ where $\alpha_i \in K$ $(i = 1, 2, \ldots, n)$ are uniquely determined. The n-tuple $(\alpha_1, \alpha_2, \ldots, \alpha_n) \in V_n(K)$ are called the **co-ordinates** of v relative to the K-basis $\mathscr{B}$, sometimes denoted by $[v]_{\mathscr{B}}$.

EXAMPLE

In Example 1 above the co-ordinates of the vector $(1, -1, 1)$ relative to the $\mathbf{R}$-basis $\{v_1, v_2, v_3\}$ are $(\frac{1}{2}, \frac{3}{4}, -\frac{1}{4})$, i.e.

$$(1, -1, 1) = \tfrac{1}{2} v_1 + \tfrac{3}{4} v_2 - \tfrac{1}{4} v_3$$

We have seen in Example 1 that in general a K-space may have more than one K-basis; in that example we saw that both K-bases have the same number of elements. The next theorem shows that this is always true. We prove

THEOREM 3.14 *Every K-basis of a finitely generated K-space has the same number of elements.*

PROOF We first show that if V is a K-space generated by a finite set of vectors $\{v_1, \ldots, v_n\}$, and if $\{y_1, y_2, \ldots, y_m\}$ is a linearly independent set of vectors in V then $m \leqslant n$.

Suppose that $m > n$, we show that this leads to a contradiction of the linear independence of $\{y_1, y_2, \ldots, y_m\}$. Since $y_j \in V$ $(j = 1, \ldots, m)$, then

$$y_j = \sum_{i=1}^{n} \alpha_{ij} v_i$$

where $\alpha_{ij} \in K$. Then, if $\beta_j \in K$ $(j = 1, \ldots, m)$

$$\sum_{j=1}^{m} \beta_j y_j = \sum_{j=1}^{m} \beta_j \left(\sum_{i=1}^{n} \alpha_{ij} v_i \right)$$

i.e. $$= \sum_{i=1}^{n} \left(\sum_{j=1}^{m} \alpha_{ij} \beta_j \right) v_i$$

Since $n < m$, by Corollary 3 to Theorem 1.7, there exist

$\beta_1, \ldots, \beta_m \in K$ (not all zero) such that $\sum\limits_{j=1}^{m} \alpha_{ij} \beta_j = 0$ which in turn

implies that $\sum\limits_{j=1}^{m} \beta_j y_j = 0$ and $\{y_1, \ldots, y_m\}$ is linearly dependent over

K. This is our required contradiction.

Now assume that $\{v_1, \ldots, v_n\}$ and $\{y_1, \ldots, y_m\}$ are K-bases of V.

By regarding $\{y_1, \ldots, y_m\}$ as a linearly independent subset in the space generated by $\{v_1, \ldots, v_n\}$ the above argument shows that $m \leqslant n$. Similarly, by regarding $\{v_1, \ldots, v_n\}$ as a linearly independent subset in the space generated by $\{y_1, \ldots, y_m\}$ it follows that $n \leqslant m$. These two results together imply that $m = n$, as required. ∎

This theorem now allows us to make the following important definition.

DEFINITION 3.15   *The number of elements in a K-basis for a finitely generated K-space V is called the **dimension** over K of V denoted by dim V or $(V : K)$. V is called a **finite dimensional vector space**.*

EXAMPLES

By scrutinizing the examples preceding the above theorem we see that

  (i) $(V_3(\mathbf{R}) : \mathbf{R}) = 3$
  (ii) $(V_n(K) : K) = n$
  (iii) $(V_n(\mathbf{C}) : \mathbf{R}) = 2n$
  (iv) $(M_n(K) : K) = n^2$
  (v) $(M_n(\mathbf{C}) : \mathbf{R}) = 2n^2$.

Our next theorem will show that a K-basis exists and at the same time gives a practical method for obtaining a K-basis V. We first prove

LEMMA 3.16   *If $\{v_1, \ldots, v_n\}$ is a linearly dependent set over K in a K-space then at least one of the v_i's is a linear combination of the vectors preceding it.*

83

PROOF If $\{v_1, \ldots, v_n\}$ is linearly dependent over K, then there exist $\alpha_i \in K$ $(i = 1, \ldots, n)$, not all zero, such that

$$\sum_{i=1}^{n} \alpha_i v_i = 0$$

Let j be the positive integer such that $1 \leqslant j \leqslant n$ and $\alpha_j \neq 0$ and $\alpha_{j+1} = \alpha_{j+2} = \ldots = \alpha_n = 0$, then

$$v_j = -\frac{\alpha_1}{\alpha_j} v_1 - \ldots - \frac{\alpha_{j-1}}{\alpha_j} v_{j-1}$$

and the lemma is proved. ∎

THEOREM 3.17 *Let V be a finitely generated K-space, then (i) every generating set for V contains a K-basis for V, and (ii) every linearly independent subset of V can be extended to give a K-basis for V, i.e. if $\{y_1, \ldots, y_m\}$ is linearly independent over K, there exist vectors $x_1, \ldots, x_{n-m}$ in V such that $\{y_1, \ldots, y_m, x_1, \ldots, x_{n-m}\}$ is a K-basis for V.*

PROOF (i) Let $S = \{x_1, \ldots, x_m\}$ generate V over K. If S is linearly independent over K then S is a K-basis for V. If not, then by Lemma 3.16, for some $1 \leqslant j \leqslant m$, x_j is a linear combination of the vectors preceding it. If $S' = S \setminus \{x_j\}$, then S' also generates V over K. If S' is linearly independent over K then this is our required K-basis. If not, repeat the above procedure on the set S' and indeed, if necessary, repeat the procedure until a linearly independent set is obtained. Such a linearly independent set exists since, for example, any set containing one vector is linearly independent.

(ii) Since V is finitely generated, suppose that $\{v_1, \ldots, v_n\}$ generates V over K. Let $\{y_1, \ldots, y_m\}$ be a linearly independent set of vectors in V, then by the proof of Theorem 3.14, $m \leqslant n$.

Consider the set

$$S = \{y_1, \ldots, y_m, v_1, \ldots, v_n\}$$

then S generates V. If S is linearly independent over K then it is the required K-basis. If not, then by Lemma 3.16, one of these vectors is linearly dependent on the vectors preceding it. This must be one of the v_i, otherwise, if one of the y_i is a linear combination of the vectors preceding it, we have a contradiction of the linear independence of $\{y_1, \ldots, y_m\}$; let this vector be v_j. Let $S' = S \setminus \{v_j\}$, then clearly S' generates V over K. If S' is linearly independent over K, we have the required K-basis for V. If not repeat the above procedure on the set S', again one of the remaining v_i must be a linear combination of the vectors preceding it. Since by (i) every generating set for V contains a K-basis for V, this process must eventually terminate and the resulting set is our required K-basis which contains $\{y_1, \ldots, y_m\}$. ∎

The following corollary is easily obtained.

COROLLARY 1 *Let V be a finite dimensional K-space with $(V:K) = n$, then (i) any set of n linearly independent vectors in V is a K-basis for V; (ii) any set of n vectors which generate V is a K-basis for V.*

PROOF (i) By the above theorem, every linearly independent set of n vectors can be extended to give a K-basis for V and since $(V:K) = n$ it follows that this set must be a K-basis for V.

(ii) Again, by the above theorem, every set of vectors which generates V contains a K-basis for V and since $(V:K) = n$ this must be our required K-basis for V. ∎

This corollary shows that if $(V:K) = n$, to verify that a set of vectors $S = \{v_1, \ldots, v_n\}$ is a K-basis for V we need only show that either

(i) S is linearly independent over K

or

(ii) S generates V over K.

We illustrate this theorem and its proof by considering the following example.

EXAMPLE

Let V be the subspace of $V_4(\mathbf{R})$ generated by $\{v_1 = (1, 1, 2, 4)$, $v_2 = (2, -1, -5, 2), v_3 = (1, -1, -4, 0), v_4 = (2, 1, 1, 6)\}$. Find (i) an $\mathbf{R}$-basis for V, (ii) an $\mathbf{R}$-basis for $V_4(\mathbf{R})$ which contains this $\mathbf{R}$-basis for V.

(i) Consider

$$\alpha_1 v_1 + \alpha_2 v_2 + \alpha_3 v_3 + \alpha_4 v_4 = 0$$

or equivalently

$$\left.\begin{aligned}
\alpha_1 + 2\alpha_2 + \alpha_3 + 2\alpha_4 &= 0 \\
\alpha_1 - \alpha_2 - \alpha_3 + \alpha_4 &= 0 \\
2\alpha_1 - 5\alpha_2 - 4\alpha_3 + \alpha_4 &= 0 \\
4\alpha_1 + 2\alpha_2 \qquad\quad + 6\alpha_4 &= 0
\end{aligned}\right\}$$

We have

$$\begin{pmatrix} 1 & 2 & 1 & 2 \\ 1 & -1 & -1 & 1 \\ 2 & -5 & -4 & 1 \\ 4 & 2 & 0 & 6 \end{pmatrix} \rightarrow \begin{pmatrix} 1 & 2 & 1 & 2 \\ 0 & -3 & -2 & -1 \\ 0 & -9 & -6 & -3 \\ 0 & -6 & -4 & -2 \end{pmatrix} \rightarrow \begin{pmatrix} 1 & 2 & 1 & 2 \\ 0 & 3 & 2 & 1 \\ 0 & 0 & 0 & 0 \\ 0 & 0 & 0 & 0 \end{pmatrix}$$

$$\rightarrow \begin{pmatrix} 1 & 0 & -\frac{1}{3} & \frac{4}{3} \\ 0 & 1 & \frac{2}{3} & \frac{1}{3} \\ 0 & 0 & 0 & 0 \\ 0 & 0 & 0 & 0 \end{pmatrix}$$

which implies that

$$\alpha_1 = \tfrac{1}{3}\alpha_3 - \tfrac{4}{3}\alpha_4$$

$$\alpha_2 = -\tfrac{2}{3}\alpha_3 - \tfrac{1}{3}\alpha_4$$

If we put $\alpha_3 = -1$, $\alpha_4 = 0$ and $\alpha_3 = 0$, $\alpha_4 = -1$ respectively we obtain,

$$v_3 = -\tfrac{1}{3}v_1 + \tfrac{2}{3}v_2$$

$$v_4 = \tfrac{4}{3}v_1 + \tfrac{1}{3}v_2$$

Thus $V = \langle v_1, v_2, v_3, v_4 \rangle = \langle v_1, v_2 \rangle$ and $\{v_1, v_2\}$ is an **R**-basis for V.

(ii) Let $S = \{v_1, v_2, e_1, e_2, e_3, e_4\}$, where $\{e_1, e_2, e_3, e_4\}$ is the standard **R**-basis for $V_4(\mathbf{R})$.
Since

$$v_1 = (1, 1, 2, 4) = e_1 + e_2 + 2e_3 + 4e_4$$

$$v_2 = (2, -1, -5, 2) = 2e_1 - e_2 - 5e_3 + 2e_4$$

we have

$$\left. \begin{array}{l} 2e_3 + 4e_4 = v_1 - e_1 - e_2 \\ 5e_3 - 2e_4 = -v_2 + 2e_1 - e_2 \end{array} \right\}$$

or $\quad e_3 = \tfrac{1}{12}(v_1 - 2v_2 + 3e_1 - 3e_2)$

$$e_4 = \tfrac{1}{24}(5v_1 + 2v_2 - 9e_1 - 3e_2)$$

which implies that

$$V_4(\mathbf{R}) = \langle v_1, v_2, e_1, e_2, e_3, e_4 \rangle = \langle v_1, v_2, e_1, e_2 \rangle$$

86

(Alternatively, using the above corollary, we need only show that $\{v_1, v_2, e_1, e_2\}$ is linearly independent over **R**).

We now proceed to obtain a useful theorem which connects the dimensions of subspaces V_1 and V_2 of a K-space V with the dimensions of the subspaces $V_1 + V_2$ and $V_1 \cap V_2$. We first need to establish,

LEMMA 3.18 *Every subspace of a finite dimensional K-space V is finite dimensional over K of dimension $\leqslant (V:K)$.*

PROOF Let W be a subspace of a K-space V. Let S be a linearly independent subset of W. Then S may be regarded as a subset of V and since V is a finite dimensional vector space, it follows from the proof of Theorem 3.14 that S contains a finite number of elements and so W is finite dimensional over K and $(W:K) \leqslant (V:K)$. ∎

THEOREM 3.19 *Let W_1 and W_2 be subspaces of a finite dimensional K-space V, then*

$$(W_1:K) + (W_2:K) = (W_1 \cap W_2:K) + (W_1 + W_2:K)$$

PROOF By Lemma 3.8, $W_1 \cap W_2$ is a subspace of V which by Lemma 3.18 is finite dimensional over K. If $(W_1 \cap W_2:K) = r < (V:K)$, let $\{v_1, \ldots, v_r\}$ be a K-basis for $W_1 \cap W_2$. Since $W_1 \cap W_2$ is a subspace of W_1 and W_2, by Theorem 3.17(ii), $\{v_1, \ldots, v_r\}$ can be extended to give K-bases for W_1 and W_2 respectively; suppose that $\{v_1, \ldots, v_r, x_1, \ldots, x_n\}$ and $\{v_1, \ldots, v_r, y_1, \ldots, y_m\}$ are K-bases for W_1 and W_2 respectively, then $(W_1:K) = r + n, (W_2:K) = r + m$. We show that $S = \{v_1, \ldots, v_r, x_1, \ldots, x_n, y_1, \ldots, y_m\}$ is a K-basis for $W_1 + W_2$, and $(W_1 + W_2:K) = r + n + m$ and this will complete the proof.

We certainly have that S generates $W_1 + W_2$ over K. We need only show that S is linearly independent over K. Consider

$$\sum_{i=1}^{r} \alpha_i v_i + \sum_{j=1}^{n} \beta_j x_j + \sum_{k=1}^{m} \gamma_k y_k = 0$$

Then we have

$$\sum_{i=1}^{r} \alpha_i v_i + \sum_{j=1}^{n} \beta_j x_j = -\sum_{k=1}^{m} \gamma_k y_k \in W_1 \cap W_2$$

Since $\{v_1, \ldots, v_r\}$ is a K-basis for $W_1 \cap W_2$, there exist $\delta_i \in K \, (i = 1, \ldots, r)$ such that

$$\sum_{k=1}^{m} \gamma_k y_k = \sum_{i=1}^{r} \delta_i v_i = 0$$

87

Since now $\{v_1, \ldots, v_r, y_1, \ldots, y_m\}$ is a K-basis for W_2 and is linearly independent over K, we have $\gamma_1 = \ldots = \gamma_m = \delta_1 = \ldots = \delta_r = 0$. Similarly, it can be proved that

$$\beta_1 = \beta_2 = \ldots = \beta_n = 0$$

and we have

$$\sum_{i=1}^{r} \alpha_i v_i = 0$$

which implies that $\alpha_i = 0$ $(i = 1, \ldots, r)$ since $\{v_1, \ldots, v_r\}$ is a K-basis for $W_1 \cap W_2$. That is

$$\alpha_i = 0 (i = 1, \ldots, r), \beta_j = 0 (j = 1, \ldots, n),$$
$$\gamma_k = 0 (k = 1, \ldots, m)$$

which proves that S is linearly independent over K and is a K-basis for $W_1 + W_2$. ∎

This theorem is illustrated by the following example:

EXAMPLE

Let W_1 and W_2 be the following subspaces of $V_4(\mathbf{R})$,
$W_1 = \{(\alpha, \beta, \gamma, \delta) \,|\, \alpha = \beta\}$, $W_2 = \{(\alpha, \beta, \gamma, \delta) \,|\, \alpha + \beta = \gamma, \delta = 2\beta\}$.
Then it is easily verified that
$\{(1, 1, 0, 0), (0, 0, 1, 0), (0, 0, 0, 1)\}$ is an $\mathbf{R}$-basis for W_1, and
$\{(1, 0, 1, 0), (0, 1, 1, 2)\}$ is an $\mathbf{R}$-basis for W_2.
$\{(1, 1, 0, 0), (0, 0, 1, 0), (0, 0, 0, 1), (1, 0, 1, 0), (0, 1, 1, 2)\}$ generates $W_1 + W_2$ over $\mathbf{R}$, but since
$(0, 1, 1, 2) = 2 (0, 0, 1, 0) + 2 (0, 0, 0, 1) + (1, 1, 0, 0) - (1, 0, 1, 0)$
and $\{(1, 1, 0, 0), (0, 0, 1, 0), (0, 0, 0, 1), (1, 0, 1, 0)\}$ is also
linearly independent over $\mathbf{R}$ it is also an $\mathbf{R}$-basis for $W_1 + W_2$.
If $(\alpha, \beta, \gamma, \delta) \in W_1 \cap W_2$, then

$$\left.\begin{array}{r} \alpha = \beta \\ 2\beta = \delta \\ \alpha + \beta = \gamma \end{array}\right\} \quad \text{or} \quad \left.\begin{array}{r} \beta = \alpha \\ \gamma = 2\alpha \\ \delta = 2\alpha \end{array}\right\}$$

and $W_1 \cap W_2 = \{(\alpha, \alpha, 2\alpha, 2\alpha) \,|\, \alpha \in \mathbf{R}\}$ and clearly $\{(1, 1, 2, 2)\}$ is an $\mathbf{R}$-basis of $W_1 \cap W_2$. We note that $(W_1 : \mathbf{R}) = 3$, $(W_2 : \mathbf{R}) = 2$, $(W_1 + W_2 : \mathbf{R}) = 4$ and $(W_1 \cap W_2 : \mathbf{R}) = 1$, which shows that

$$(W_1 : \mathbf{R}) + (W_2 : \mathbf{R}) = (W_1 + W_2 : \mathbf{R}) + (W_1 \cap W_2 : \mathbf{R})$$

Exercises 3.4

1. Prove that
   (i) $\{(1, 0, 3), (5, 2, 1), (0, 1, 6)\}$ is an **R**-basis for $V_3(\mathbf{R})$,
   (ii) if t is a fixed real number and $g_1(x) = 1, g_2(x) = x + t$,
$g_3(x) = (x + t)^2$ then $\{g_1, g_2, g_3\}$ is an **R**-basis for $P_2(\mathbf{R})$.

2. Prove that $\{(1, 2, 0), (0, 5, 7), (-1, 1, 3)\}$ is an **R**-basis for $V_3(\mathbf{R})$
and find the co-ordinates of $(0, 13, 17)$ and $(2, 3, 1)$ relative to this
R-basis.

3. (i) Show that $\{(3 + \sqrt{2}, 1 + \sqrt{2}), (7, 1 + 2\sqrt{2})\}$ is linearly
dependent over **R** but linearly independent over **Q**.
   (ii) Show that $\{(1 - i, i), (2, -1 + i)\}$ is linearly dependent over **C**
but is linearly independent over **R**.

4. Show that (i) $\{e^t, \sin t, t^2\}$ (ii) $\{e^t, \sin t, \cos t\}$ are linearly independent
over **R**.

5. Show that the subset $\{(1, 0, 1, 1), (1, 0, 2, 4)\}$ of $V_4(\mathbf{R})$ is linearly
independent over **R** and extend it to an **R**-basis for $V_4(\mathbf{R})$.

6. If $\{u, v, w\}$ is linearly independent over **C** in a **C**-space V, prove that
   (i) $\{u + v, v + w, w + u\}$ is linearly independent over **C**,
   (ii) $\{u + v - 3w, u + 3v - w, v + w\}$ is linearly dependent over **C**.

7. Prove that $\{(1, 1, 0, -1), (4, -2, 1, 0)\}$ is linearly independent over
Q and find a **Q**-basis for $V_4(\mathbf{Q})$ containing these two vectors.

8. Let $M = \left\{ \begin{pmatrix} a & b \\ -b & c \end{pmatrix} \mid a, b, c \in \mathbf{R} \right\}$ and $N = \left\{ \begin{pmatrix} x & 0 \\ y & 0 \end{pmatrix} \mid x, y \in \mathbf{R} \right\}$
be subspaces of $M_2(\mathbf{R})$. Find **R**-bases for $M, N, M \cap N$ and $M + N$.

9. If S and T are subspaces of $V_4(\mathbf{R})$ defined by
$S = \{(\alpha, \beta, \gamma, \delta) \mid \alpha + \beta + \gamma = 0\}$ and $T = \{(\alpha, \beta, \gamma, \delta) \mid \gamma = -\delta\}$,
find **R**-bases for $S + T$ and $S \cap T$.

10. Find an **R**-basis for the solution space of the equation

$$x - 2y + z - 3t = 0$$

11. If V_1 and V_2 are 2-dimensional subspaces in $V_3(\mathbf{R})$, prove that
$(V_1 \cap V_2 : \mathbf{R}) > 0$.

12. Let U be the subspace of $V_3(\mathbf{R})$ generated by the two vectors
$u_1 = (1, 2, 3), u_2 = (3, -5, 1)$. Show that $(1, 0, 0)$ is not in U but that
$(5, -23, -9)$ is in U. Express the latter vector as a linear combination
of u_1 and u_2.

89

13. Determine whether or not each of the following is an **R**-basis for $P_n(\mathbf{R})$

(i) $\{1, 1 + t, 1 + t + t^2, \ldots, 1 + t + \ldots + t^n\}$,

(ii) $\{1 + t, t + t^2, \ldots, t^{n-1} + t^n\}$,

(iii) $\{1, 1 - t, \ldots, (1 - t)^n\}$.

14. Are the vectors $u = (3, -1, 0, -1)$ and $v = (1, 0, 4, -1)$ in the subspace of $V_4(\mathbf{R})$ generated by $\{(2, -1, 3, 2), (-1, 1, 1, -3),$ $(1, 1, 9, -5)\}$? Hence determine two **R**-bases of $V_4(\mathbf{R})$ one containing u and one containing v.

15. If S is a linearly independent set of vectors in V and $v \in V$, prove that $v \notin \langle S \rangle$ if and only if $S \cup \{v\}$ is linearly independent.

Show that the set $S = \{(1, 0, -1, 1), (2, -1, 0, 1), (1, 1, 2, 1)\}$ is linearly independent over **R** and that $x = (1, 3, 3, 2) \in \langle S \rangle$ and $y = (0, 1, 1, -1) \notin \langle S \rangle$. Find an **R**-basis for $\langle S \rangle$ which contains x and a **R**-basis for $V_4(\mathbf{R})$ which contains y.

16. Show that $M_2(\mathbf{C})$ may be regarded as a **C**-space and an **R**-space and determine its dimension in each case. In each case, determine the dimension of the subspace generated by

$$\left\{ \begin{pmatrix} 1 & 0 \\ 0 & 1 \end{pmatrix}, \begin{pmatrix} 1 & 0 \\ 0 & i \end{pmatrix}, \begin{pmatrix} i & 0 \\ 0 & 1 \end{pmatrix}, \begin{pmatrix} i & 0 \\ 0 & i \end{pmatrix} \right\}.$$

17. If S and T are subspaces of $V_3(\mathbf{C})$ generated by $\{(1, 1, 0), (i, 1 + i, 1), (1 + i, 1 + i, 0)\}$ and $\{(1, 0, 1), (i, -i, 0), (0, i, i)\}$ respectively, find **C**-bases for $S \cap T$ and $S + T$.

Regarding the above as real spaces, show that if S' and T' are the real subspaces generated by the above sets, prove that $S' \cap T' = \{0\}$.

18. Let V be the subspace of $P_3(\mathbf{R})$ generated by the polynomials $1 - t^2 + t^3, 2 + t - t^2 + t^3, 1 + 2t + t^2 - t^3$. Show that $f(t) = t + t^2 - t^3 \in V$ but $g(t) = 1 + t - t^2 + t^3 \notin V$. Find an **R**-basis for V which contains $f(t)$ and an **R**-basis for $P_3(\mathbf{R})$ which contains $g(t)$.

CHAPTER 4

Linear Transformations on Vector Spaces

4.1 Linear Transformations

Let V and W be K-spaces.

DEFINITION 4.1   *A **linear transformation** (**K-homomorphism**) T from V into W is a mapping $T : V \to W$ such that*

$$T(\alpha v + v') = \alpha T(v) + T(v')$$

for all $\alpha \in K$, v, $v' \in V$, or equivalently

$$T(v + v') = T(v) + T(v')$$

$$T(\alpha v) = \alpha T(v)$$

*for all $\alpha \in K$, v, $v' \in V$. If $W = V$ then we say that T is a **linear transformation on V**.*

Essentially, a mapping from V into W is a linear transformation if it respects the two basic operations in a vector space, namely addition and scalar multiplication by elements of K, as illustrated in the following diagram.

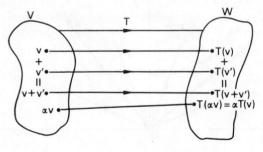

Figure 1

91

The following examples illustrate the breadth of the concept of a linear transformation— the first two show that it is a generalization to arbitrary vector spaces of some well known "transformations" in the plane $V_2(\mathbf{R})$, namely reflection in a line and rotation about a point.

EXAMPLES

1. Let L be a line through the origin and let T be the mapping which reflects each vector in L

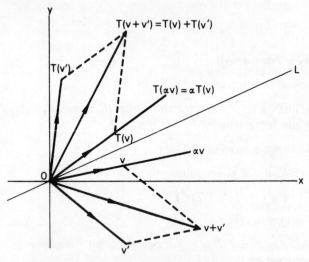

Figure 2

2. Let R be a rotation through an angle θ about the origin

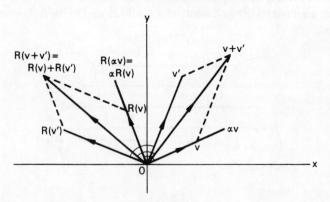

Figure 3

92

3. Let $T : V_3(\mathbf{R}) \to V_2(\mathbf{R})$ be defined by

$$T(\alpha_1, \alpha_2, \alpha_3) = (\alpha_1 + \alpha_2, \alpha_2 - \alpha_3)$$

then T is a linear transformation of $V_3(\mathbf{R})$ into $V_2(\mathbf{R})$ since if
$(\alpha_1, \alpha_2, \alpha_3), (\beta_1, \beta_2, \beta_3) \in V_3(\mathbf{R}), \alpha \in \mathbf{R}$ then

$$T(\alpha(\alpha_1, \alpha_2, \alpha_3) + (\beta_1, \beta_2, \beta_3)) = T(\alpha\alpha_1 + \beta_1, \alpha\alpha_2 + \beta_2, \alpha\alpha_3 + \beta_3)$$

$$= (\alpha\alpha_1 + \beta_1 + \alpha\alpha_2 + \beta_2, \alpha\alpha_2 + \beta_2 - \alpha\alpha_3 - \beta_3)$$

$$= \alpha(\alpha_1 + \alpha_2, \alpha_2 - \alpha_3) + (\beta_1 + \beta_2, \beta_2 - \beta_3)$$

$$= \alpha T(\alpha_1, \alpha_2, \alpha_3) + T(\beta_1, \beta_2, \beta_3)$$

4. Let $T : V_3(\mathbf{R}) \to V_2(\mathbf{R})$ be defined by

$$T(\alpha_1, \alpha_2, \alpha_3) = (\alpha_1 + 1, \alpha_2)$$

then T is **not** a linear transformation since for example $T(1, 0, 0) = (2, 0)$
and $T(2(1, 0, 0)) = T(2, 0, 0) = (3, 0)$ and
$T(2(1, 0, 0)) \neq 2T(1, 0, 0) = (4, 0)$.

5. Let $D : P_n(\mathbf{R}) \to P_n(\mathbf{R})$ be the differentiation mapping defined by mapping
a polynomial onto its derivative

$$D(\alpha_0 + \alpha_1 x + \ldots + \alpha_n x^n) = \alpha_1 + 2\alpha_2 x + 3\alpha_3 x^2 + \ldots + n\alpha_n x^{n-1}$$

then D is a linear transformation on $P_n(\mathbf{R})$ since if
$\alpha, \alpha_i, \beta_i \in \mathbf{R}$ $(i = 1, \ldots, n)$

$$D(\alpha(\alpha_0 + \alpha_1 x + \ldots + \alpha_n x^n) + (\beta_0 + \beta_1 x + \ldots + \beta_n x^n))$$
$$= D((\alpha\alpha_0 + \beta_0) + (\alpha\alpha_1 + \beta_1)x + \ldots + (\alpha\alpha_n + \beta_n)x^n)$$
$$= (\alpha\alpha_1 + \beta_1) + 2(\alpha\alpha_2 + \beta_2)x + \ldots + n(\alpha\alpha_n + \beta_n)x^{n-1}$$
$$= \alpha D(\alpha_0 + \alpha_1 x + \ldots + \alpha_n x^n) + D(\beta_0 + \beta_1 x + \ldots + \beta_n x^n)$$

6. More generally, let D map a continuously differentiable function on
$[a,b]$ onto its derivative, that is $D : C'[a,b] \to C[a,b]$ is the derivative
map. Then D is a linear transformation since if $\alpha \in \mathbf{R}, f_1, f_2 \in C'[a,b]$,
then

$$D(f_1 + f_2) = D(f_1) + D(f_2)$$

$$D(\alpha f_1) = \alpha D(f_1)$$

are restatements of well known properties of determinants (see A.S.-T. Lue,
Basic Pure Mathematics II, VNR New Mathematics Library 5, p.56).

7. Integration of functions is also a linear transformation; more precisely, if $f \in C[a,b]$ define

$$S(f) = \int_a^x f(t) \, dt$$

where $a \leqslant x \leqslant b$. Then $S : C[a,b] \to C[a,b]$ and

$$S(f_1 + f_2) = \int_a^x (f_1(t) + f_2(t)) dt$$

$$= \int_a^x f_1(t) dt + \int_a^x f_2(t) dt$$

$$= S(f_1) + S(f_2)$$

and similarly

$$S(\alpha f_1) = \alpha S(f_1)$$

for $f_1, f_2 \in C[a,b]$, $\alpha \in \mathbf{R}$.

8. Let M and N be fixed $m \times m$ and $n \times n$ matrices respectively. Define $T : M_{m,n}(K) \to M_{m,n}(K)$ by $T(A) = MAN$ for all $A \in M_{m,n}(K)$, then if $A, B \in M_{m,n}(K)$, and $\alpha \in K$ we have

$$T(\alpha A + B) = M(\alpha A + B)N$$

$$= \alpha MAN + MBN$$

$$= \alpha T(A) + T(B)$$

i.e. T is a linear transformation on $M_{m,n}(K)$

9. If V and W are K-spaces, then $I : V \to V$ defined by $I(v) = v$ for all $v \in V$ is a linear transformation on V called the **identity** transformation and $0 : V \to W$ defined by $0(v) = 0$ for all $v \in V$ is a linear transformation from V into W called the **zero** transformation.

Remark 1 $T(0) = 0$ since by Theorem 3.2 we have $0 + 0 = 0$ which implies $T(0) + T(0) = T(0)$ and $T(0) + 0 = T(0)$ and the result follows from the uniqueness of the zero element (Theorem 3.4).

Remark 2   If $v_i \in V$, $\alpha_i \in K$, $(i = 1, \ldots, n)$, then

$$T \left(\sum_{i=1}^n \alpha_i v_i \right) = \sum_{i=1}^n \alpha_i T(v_i)$$

This result follows by repeated application of the definition of a linear transformation.

We now prove a theorem which will be useful later when the existence of a linear transformation with certain properties is required.

94

THEOREM 4.2   *Let V and W be K-spaces where $(V : K) = n < \infty$.*
If $\{v_1, \ldots, v_n\}$ is a K-basis for V and $\{w_1, \ldots, w_n\}$ are any n vectors
in W, then there exists a unique linear transformation $T : V \to W$ such
that $T(v_i) = w_i$ $(i = 1, 2, \ldots, n)$.

PROOF   If $v \in V$, then $v = \sum\limits_{i=1}^{n} \alpha_i v_i$, where $\alpha_i \in K$ $(i = 1, \ldots, n)$ are

uniquely determined. Define $T : V \to W$ by

$$T(v) = T\left(\sum_{i=1}^{n} \alpha_i v_i \right) = \sum_{i=1}^{n} \alpha_i w_i$$

If $v, v' \in V$, $\alpha \in K$ and $v = \sum\limits_{i=1}^{n} \alpha_i v_i$, $v' = \sum\limits_{i=1}^{n} \beta_i v_i$, where

$\alpha_i, \beta_i \in K$ $(i = 1, \ldots, n)$ then

$$T(\alpha v + v') = T\left(\sum_{i=1}^{n} (\alpha \alpha_i + \beta_i) \, v_i \right)$$

$$= \sum_{i=1}^{n} (\alpha \alpha_i + \beta_i) w_i$$

$$= \alpha T(v) + T(v')$$

and so T is a linear transformation of V into W, which clearly has the
property that $T(v_i) = w_i$ $(i = 1, \ldots, n)$. If $S : V \to W$ is a linear
transformation with the same property $S(v_i) = w_i$ $(i = 1, \ldots, n)$ then
if $v \in V$ and $v = \sum\limits_{i=1}^{n} \alpha_i v_i$, $\alpha_i \in K$ $(i = 1, \ldots, n)$ then

$$S(v) = S\left(\sum_{i=1}^{n} \alpha_i v_i \right) = \sum_{i=1}^{n} \alpha_i S(v_i) = \sum_{i=1}^{n} \alpha_i w_i = T(v)$$

and $S = T$, which proves the uniqueness of T. ∎

Exercises 4.1

1. Determine whether the following mappings $T : V_3(\mathbf{R}) \to V_2(\mathbf{R})$ are
linear transformations

   (i) $T(\alpha, \beta, \gamma) = (\alpha + \beta - \gamma, 2\alpha + \beta)$
   (ii) $T(\alpha, \beta, \gamma) = (\alpha + 1, \alpha + 2\beta - \gamma)$
   (iii) $T(\alpha, \beta, \gamma) = (|\alpha|, 0)$
   (iv) $T(\alpha, \beta, \gamma) = (\alpha\beta, \beta\alpha)$.

2. Determine whether the following mappings $T : M_n(K) \to M_n(K)$ are linear transformations

(i) $T(A) = AS$, where S is a fixed matrix in $M_n(K)$
(ii) $T(A) = AS - SA$, where S is a fixed matrix in $M_n(K)$
(iii) $T(A) = A^t$
(iv) $T(A) = A^2$.

3. Determine whether the following mappings $T : C'(\mathbf{R}) \to C(\mathbf{R})$ are linear transformations

(i) $T(f(x)) = f'(x)$

(ii) $T(f(x)) = \int_0^x f(t)\, dt$

(iii) $T(f(x)) = f(x) f'(x)$
(iv) $T(f(x)) = xf(x)$
(v) $T(f(x)) = f(x + 1)$.

4. (i) Define $T : C^2(\mathbf{R}) \to C(\mathbf{R})$ by $T(f(x)) = f''(x) - 2f'(x) + 3$. Show that T is a linear transformation;

(ii) If $a(x), b(x) \in C(\mathbf{R})$, define $T : C^2(\mathbf{R}) \to C(\mathbf{R})$ by
$T(f(x)) = f''(x) + a(x) f'(x) + b(x) f(x)$,
and show that T is a linear transformation.

4.2 The Matrix of a Linear Transformation

Let V and W be finite dimensional vector spaces over K, where $(V : K) = n$, $(W : K) = m$. Let $\mathscr{B} = \{v_1, \ldots, v_n\}$ and $\mathscr{W} = \{w_1, \ldots, w_m\}$ be K-bases for V and W respectively. For $j = 1, \ldots, n$, $Tv_j \in W$ and thus

$$Tv_j = \sum_{i=1}^{m} \alpha_{ij} w_i$$

where $\alpha_{ij} \in K$ are uniquely determined. Let $A = (\alpha_{ij}) \in M_{m,n}(K)$, then A is called the **matrix of T relative to the K-bases $\mathscr{B}$ and $\mathscr{W}$** and is sometimes written $_{\mathscr{B}}(T)_{\mathscr{W}}$. Note that the coefficients involved in Tv_j give the elements of the jth column of A.

Conversely, if $A = (\alpha_{ij}) \in M_{m,n}(K)$, define $T_A : V \to W$ by

$$T_A v_j = \sum_{i=1}^{n} \alpha_{ij} w_i \qquad (j = 1, \ldots, n)$$

then by Theorem 4.2, T_A is the unique linear transformation with this property. Thus, there is a one-one correspondence between the set of linear transformations from V into W and the set of all $m \times n$ matrices over K.

If T is a linear transformation on V, we denote $_\mathscr{B}(T)_\mathscr{B}$ by $(T)_\mathscr{B}$.

We consider in particular a linear transformation T from $V_n(K)$ into $V_m(K)$. If $(x_1, x_2, \ldots, x_n) \in V_n(K)$, then put $T(x_1, x_2, \ldots, x_n)$ $= (y_1, y_2, \ldots, y_m) \in V_m(K)$. If $\mathscr{B} = \{e_1, e_2, \ldots, e_n\}$ and $\mathscr{W}$ $= \{e_1', e_2', \ldots, e_m'\}$ are the standard K-bases for $V_n(K)$ and $V_m(K)$ respectively and $_\mathscr{B}(T)_\mathscr{W} = (\alpha_{ij})$, then for $j = 1, \ldots, n$

$$T(e_j) = \sum_{i=1}^{m} \alpha_{ij} e_i'$$

$$= (\alpha_{1j}, \alpha_{2j}, \ldots, \alpha_{mj})$$

Thus, we have

$$(y_1, y_2, \ldots, y_m) = T(x_1, x_2, \ldots, x_n)$$

$$= \sum_{j=1}^{n} x_j T(e_j)$$

$$= \sum_{j=1}^{n} x_j (\alpha_{1j}, \alpha_{2j}, \ldots, \alpha_{mj})$$

or in other words

$$y_i = \sum_{j=1}^{n} \alpha_{ij} x_j \qquad (i = 1, \ldots, m)$$

This means that if $T: V_n(K) \to V_m(K)$ is a linear transformation and $A = (\alpha_{ij}) \in M_{m,n}(K)$ is the matrix of T relative to the standard K-bases for $V_n(K)$ and $V_m(K)$ then

$$T(x_1, x_2, \ldots, x_n) = \left(\sum_{j=1}^{n} \alpha_{1j} x_j, \sum_{j=1}^{n} \alpha_{2j} x_j, \ldots, \sum_{j=1}^{n} \alpha_{mj} x_j \right) \quad (1)$$

Conversely, if $T: V_n(K) \to V_m(K)$ is defined by (1), then it is easily verified that T is a linear transformation. Hence every linear transformation from $V_n(K)$ into $V_m(K)$ must be of this form. Note that if the linear transformation T is presented in this form, the ith component of $T(x_1, \ldots, x_n)$ gives the elements in the ith row of the matrix of T relative to the standard K-basis for $V_n(K)$ and $V_m(K)$

EXAMPLES

1. Let $T : V_3(\mathbf{R}) \to V_2(\mathbf{R})$ be defined by $T(\alpha, \beta, \gamma) = (\alpha + \beta - \gamma, 2\alpha + \gamma)$ for all $(\alpha, \beta, \gamma) \in V_3(\mathbf{R})$. Then, by the above, T is a linear transformation and the matrix of T relative to the standard $\mathbf{R}$-bases for $V_3(\mathbf{R})$ and $V_2(\mathbf{R})$ is

$$\begin{pmatrix} 1 & 1 & -1 \\ 2 & 0 & 1 \end{pmatrix}$$

Now, let $\mathscr{B} = \{v_1 = (1, 0, -1), v_2 = (1, 1, 1), v_3 = (1, 0, 0)\}$ and $\mathscr{W} = \{w_1 = (1, 1), w_2 = (1, 0)\}$ then it is easily verified that $\mathscr{B}$ and $\mathscr{W}$ are linearly independent over $\mathbf{R}$ and are $\mathbf{R}$-bases for $V_3(\mathbf{R})$ and $V_2(\mathbf{R})$ respectively. We now determine the matrix of T relative to $\mathscr{B}$ and $\mathscr{W}$. We have

$$Tv_1 = T(1, 0, -1) = (2, 1) = w_1 + w_2$$
$$Tv_2 = T(1, 1, \ \ 1) = (1, 3) = 3w_1 - 2w_2$$
$$Tv_3 = T(0, 0, \ \ 1) = (1, 2) = 2w_1 - w_2$$

and so

$$_{\mathscr{B}}(T)_{\mathscr{W}} = \begin{pmatrix} 1 & 3 & 2 \\ 1 & -2 & -1 \end{pmatrix}$$

2. Let $R : V_2(\mathbf{R}) \to V_2(\mathbf{R})$ be the reflection on the line L with angle ϕ through the origin. We find the matrix of R relative to the standard $\mathbf{R}$-bases for $V_2(\mathbf{R})$.

To determine this matrix, we must find $R(1, 0)$ and $R(0, 1)$. From Figure 4, we see that

$$R(1, 0) = (\cos 2\phi, \sin 2\phi)$$
$$R(0, 1) = (\sin 2\phi, -\cos 2\phi)$$

and the required matrix is

$$\begin{pmatrix} \cos 2\phi & \sin 2\phi \\ \sin 2\phi & -\cos 2\phi \end{pmatrix}$$

98

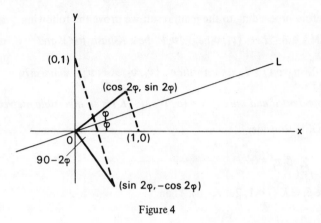

Figure 4

3. Let R be a rotation through an angle θ about the origin in $V_2(\mathbf{R})$.

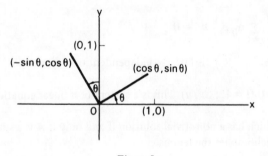

Figure 5

In this case

$$R(1, 0) = (\cos \theta, \sin \theta)$$
$$R(0, 1) = (-\sin \theta, \cos \theta)$$

and so, the matrix of R is $\begin{pmatrix} \cos \theta & -\sin \theta \\ \sin \theta & \cos \theta \end{pmatrix}$.

4.3 Change of Basis

In the first example on page 98, we saw that the matrix of a linear transformation is dependent on the choice of basis. We now investigate more closely the effect of a change of K-basis on the matrix of a linear transformation.

99

Before proceeding to the main result we prove the following:

LEMMA 4.3   Let $\{v_1, v_2, \ldots, v_n\}$ be a K-basis for V and

$$v_j' = \sum_{i=1}^{n} \alpha_{ij} \, v_i \,(j = 1, \ldots, n), \text{ then } \{v_1', v_2', \ldots, v_n'\} \text{ is linearly}$$

independent if and only if $A = (\alpha_{ij}) \in M_n (K)$ is an invertible matrix.

PROOF   Consider

$$\sum_{j=1}^{n} \beta_j v_j' = 0$$

where $\beta_j \in K \,(j = 1, 2, \ldots, n)$. Then

$$\sum_{j=1}^{n} \beta_j \left(\sum_{i=1}^{n} \alpha_{ij} \, v_i \right) = 0$$

i.e.

$$\sum_{i=1}^{n} \left(\sum_{j=1}^{n} \alpha_{ij} \beta_j \right) v_i = 0$$

But $\{v_1, v_2, \ldots, v_n\}$ is linearly independent over K, and so

$$\sum_{j=1}^{n} \alpha_{ij} \beta_j = 0 \,(i = 1, \ldots, n).$$ This is a system of n linear equations in n

variables which has a non-trivial solution if and only if A is a singular
matrix, which implies the lemma.

COROLLARY   A matrix A is invertible if and only if its columns
(rows) are linearly independent over K.

Let $T : V \to W$ be a linear transformation where $(V : K) = n$,
$(W : K) = m$. Let $\mathscr{B} = \{v_1, v_2, \ldots, v_n\}$ and $\mathscr{W} = \{w_1, w_2, \ldots, w_m\}$ be
K-bases for V and W respectively and let ${}_{\mathscr{B}}(T)_{\mathscr{W}} = A = (\alpha_{ij}) \in M_{m,n}(K)$ be
the matrix of T relative to $\mathscr{B}$ and $\mathscr{W}$, i.e.

$$Tv_j = \sum_{i=1}^{m} \alpha_{ij} w_i \qquad (j = 1, \ldots, n)$$

Now, let $\mathscr{B}' = \{v_1', v_2', \ldots, v_n'\}$ and $\mathscr{W}' = \{w_1', w_2', \ldots, w_m'\}$ also be
K-bases for V and W respectively and let ${}_{\mathscr{B}_1'}(T)_{\mathscr{W}'} = B = (\beta_{ij}) \in M_{m,n}(K)$ be
the matrix of T relative to $\mathscr{B}'$ and $\mathscr{W}'$, i.e.

$$Tv_j' = \sum_{i=1}^{m} \beta_{ij} w_i' \qquad (j = 1, 2, \ldots, n)$$

100

Now, $\mathscr{B}$ and $\mathscr{B}'$ are K-bases for V and by Theorem 4.2 there exists a unique linear transformation P on V such that $Pv_j = v_j'\,(j = 1, 2, \ldots, n)$. Similarly, there exists a unique linear transformation Q on W such that $Qw_j = w_j'\,(j = 1, 2, \ldots, m)$. Let $C = (\gamma_{ij}) \in M_n(K)$ and $D = (\delta_{ij}) \in M_m(K)$ be the matrices of P and Q relative to the K-basis $\mathscr{B}'$ and $\mathscr{W}$ respectively i.e.

$$v_j' = Pv_j = \sum_{i=1}^{n} \gamma_{ij}\, v_i \qquad (j = 1, 2, \ldots, n)$$

and

$$w_j' = Qw_j = \sum_{i=1}^{m} \delta_{ij}\, w_i \qquad (j = 1, 2, \ldots, m)$$

Then, for $j = 1, 2, \ldots, n$, we have

$$Tv_j' = \sum_{i=1}^{m} \beta_{ij}\, w_i'$$

$$= \sum_{i=1}^{m} \beta_{ij} \left(\sum_{k=1}^{m} \delta_{ki}\, w_k \right)$$

$$= \sum_{k=1}^{m} \left(\sum_{i=1}^{m} \delta_{ki}\, \beta_{ij} \right) w_k$$

and alternatively

$$Tv_j' = T \left(\sum_{i=1}^{n} \gamma_{ij}\, v_i \right)$$

$$= \sum_{i=1}^{n} \gamma_{ij} \left(\sum_{k=1}^{m} \alpha_{ki}\, w_k \right)$$

$$= \sum_{k=1}^{m} \left(\sum_{i=1}^{n} \alpha_{ki}\, \gamma_{ij} \right) w_k$$

Thus, we have two alternative forms for the matrix of T relative to the K-bases $\mathscr{B}'$ and $\mathscr{W}$, namely

$$\mathscr{B}'(T)_{\mathscr{W}} = DB = AC$$

By Lemma 4.3 proved above, we have that C and D are invertible matrices, thus

$$B = D^{-1}AC \qquad\qquad (2)$$

101

Thus, we have proved the following:

THEOREM 4.4 *Let T be a linear transformation from a K-space V into a K-space W. If $\mathscr{B}$ and $\mathscr{B}'$ are K-bases for V and $\mathscr{W}$ and $\mathscr{W}'$ are K-bases for W, then*

$$_{\mathscr{B}'}(T)_{\mathscr{W}'} = (Q)^{-1}_{\mathscr{W}} \; _{\mathscr{B}}(T)_{\mathscr{W}} \; (P)_{\mathscr{B}}$$

where P is the unique linear transformation which maps $\mathscr{B}$ onto $\mathscr{B}'$ and Q is the unique linear transformation which maps $\mathscr{W}$ onto $\mathscr{W}'$.

This theorem is illustrated by applying it to the first example considered on page 98. Using the notation of that example, then

$$
\begin{aligned}
_{\mathscr{B}'}(T)_{\mathscr{W}'} &= \begin{pmatrix} 1 & 1 \\ 1 & 0 \end{pmatrix}^{-1} \begin{pmatrix} 1 & 1 & -1 \\ 2 & 0 & 1 \end{pmatrix} \begin{pmatrix} 1 & 1 & 1 \\ 0 & 1 & 0 \\ -1 & 1 & 0 \end{pmatrix} \\[2mm]
&= \begin{pmatrix} 0 & 1 \\ 1 & -1 \end{pmatrix} \begin{pmatrix} 2 & 1 & 1 \\ 1 & 3 & 2 \end{pmatrix} \\[2mm]
&= \begin{pmatrix} 1 & 3 & 2 \\ 1 & -2 & -1 \end{pmatrix}
\end{aligned}
$$

which conforms with the result obtained there.

We now consider the particular case when T is a linear transformation on V in more detail. In this case, we let the K-bases $\mathscr{B}$ and $\mathscr{W}$ coincide and the K-bases $\mathscr{B}'$ and $\mathscr{W}'$ coincide. This means now that $D = C$ and thus

$$(T)_{\mathscr{B}'} = C^{-1} (T)_{\mathscr{B}} C$$

We give a special name for matrices connected in this way.

DEFINITION 4.5 *If $A, B \in M_n(K)$, then we say that B is **similar** to A if there exists an invertible matrix $C \in M_n(K)$ such that*

$$B = C^{-1} AC$$

In fact, what we have proved above is that if A and B represent T relative to certain K-bases for V, then A and B are similar. The converse of this can also be proved. We have

THEOREM 4.6 *Let $A, B \in M_n(K)$, then A and B are similar if and only if they represent the same linear transformation on a vector space V of dimension n relative to suitably chosen K-bases for V.*

102

PROOF   If A and B are similar matrices, then by Definition 4.5 there exists an invertible matrix $C \in M_n(K)$ such that $B = C^{-1}AC$. Let V be a K-space of dimension n with K-basis $\{v_1, v_2, \ldots, v_n\}$ and let T be the linear transformation T on V defined by

$$Tv_j = \sum_{i=1}^{n} \alpha_{ij} v_i \qquad (j = 1, \ldots, n)$$

For $j = 1, 2, \ldots, n$, let

$$v_j' = \sum_{i=1}^{n} \gamma_{ij} v_i$$

where $C = (\gamma_{ij})$ and $C^{-1} = (\gamma_{ij}')$. Then by Lemma 4.3 since C is invertible, $\{v_1', v_2', \ldots, v_n'\}$ is also a K-basis for V. Furthermore, for $j = 1, 2, \ldots, n$,

$$Tv_j' = T\left(\sum_{i=1}^{n} \gamma_{ij} v_i \right)$$

$$= \sum_{i=1}^{n} \gamma_{ij} \left(\sum_{k=1}^{n} \alpha_{ki} v_k \right)$$

$$= \sum_{k=1}^{n} \left(\sum_{i=1}^{n} \alpha_{ki} \gamma_{ij} \right) v_k$$

$$= \sum_{k=1}^{n} \left(\sum_{i=1}^{n} \alpha_{ki} \gamma_{ij} \right) \left(\sum_{\ell=1}^{n} \gamma_{\ell k}' v_\ell' \right)$$

$$= \sum_{\ell=1}^{n} \left(\sum_{k=1}^{n} \sum_{i=1}^{n} \gamma_{\ell k}' \alpha_{ki} \gamma_{ij} \right) v_\ell'$$

Thus, the matrix of T relative to the K-basis $\{v_1', v_2', \ldots, v_n'\}$ for V is $C^{-1}AC = B$. This completes the proof since the proof of the converse has been given before the statement of Theorem 4.6. ∎

EXAMPLE   Let $\mathscr{B} = \{u, v, w\}$ be a K-basis for a 3-dimensional vector space V and let

$$\begin{pmatrix} 1 & -1 & 2 \\ 2 & 0 & 1 \\ 1 & -1 & 2 \end{pmatrix}$$

be the matrix of a linear transformation on V relative to this K-basis. Find the matrix of T relative to the K-basis $\mathcal{B}' = \{u + v, u - 2v + w, v - w\}$.

By the above theorem, the matrix

$$C = \begin{pmatrix} 1 & 1 & 0 \\ 1 & -2 & 1 \\ 0 & 1 & -1 \end{pmatrix}$$

and by the methods of §1.6, we can invert this matrix to give

$$C^{-1} = \frac{1}{2} \begin{pmatrix} 1 & 1 & 1 \\ 1 & -1 & -1 \\ 1 & -1 & -3 \end{pmatrix}$$

Thus, by Theorem 4.6, the matrix of T relative to the K-basis $\mathcal{B}'$ is

$$\frac{1}{2} \begin{pmatrix} 1 & 1 & 1 \\ 1 & -1 & -1 \\ 1 & -1 & -3 \end{pmatrix} \begin{pmatrix} 1 & -1 & 2 \\ 2 & 0 & 1 \\ 1 & -1 & 2 \end{pmatrix} \begin{pmatrix} 1 & 1 & 0 \\ 1 & -2 & 1 \\ 0 & 1 & -1 \end{pmatrix}$$

$$= \frac{1}{2} \begin{pmatrix} 2 & 13 & -7 \\ -2 & -3 & 1 \\ -2 & -13 & 7 \end{pmatrix}$$

Exercises 4.3

1. Let T be a linear transformation on $V_3(\mathbf{R})$ defined by

$$T(x, y, z) = (x - y, x + 2y - z, 2x + y + z)$$

Find the matrix of T relative to (i) the standard basis for $V_3(\mathbf{R})$; (ii) the $\mathbf{R}$-basis $\{v_1, v_2, v_3\}$ for $V_3(\mathbf{R})$, where $v_1 = (1, 0, 1)$, $v_2 = (-2, 1, 1)$, $v_3 = (1, -1, 1)$.

2. The matrix of a linear transformation T on $V_3(\mathbf{R})$ relative to the standard basis is

$$\begin{pmatrix} 0 & 1 & 1 \\ 1 & 0 & 1 \\ 1 & 1 & 0 \end{pmatrix}$$

Find the matrix of T relative to the **R**-basis $\{v_1, v_2, v_3\}$ where $v_1 = (1, 0, 1)$, $v_2 = (-2, 1, 1)$, $v_3 = (1, -1, 1)$.

3. Find the matrix of the linear transformations T on P_n (**R**) defined by (a) $T(f(x)) = f'(x)$ (b) $T(f(x)) = f(x+1)$, relative to the **R**-basis
(i) $\{1, x, x^2, \ldots, x^n\}$, (ii) $\{1, x - 1, (x - 1)^2, \ldots, (x - 1)^n\}$,
(iii) $\{1, 1 + x, 1 + x + x^2, \ldots, 1 + x + \ldots + x^n\}$.

4. If S is a fixed matrix in M_2 (**R**), find the matrix of each of the following linear transformations T on M_2 (**R**) relative to the standard **R**-basis $\{e_{ij} | i, j = 1, 2)\}$
   (i) $TA = SA$,   (ii) $TA = AS$,   (iii) $TA = SA - AS$.

5. If $\{u_1, u_2\}$ and $\{v_1, v_2, v_3\}$ are **R**-bases for $V_2(\mathbf{R})$ and $V_3(\mathbf{R})$ respectively and if a linear transformation T from $V_2(\mathbf{R})$ into $V_3(\mathbf{R})$ is defined by

$$Tu_1 = v_1 + 2v_2 - v_3$$
$$Tu_2 = v_1 - v_2$$

find the matrix of T relative to these bases. Find also the matrix of T relative to the **R**-bases $\{-u_1 + u_2, 2u_1 - u_2\}$ and $\{v_1, v_1 + v_2, v_1 + v_2 + v_3\}$ for $V_2(\mathbf{R})$ and $V_3(\mathbf{R})$ respectively. What is the relationship between these two matrices?

6. If U and V are K-spaces of dimension 3 and $\mathscr{M} = \{u_1, u_2, u_3\}$,

$\mathscr{B} = \{v_1, v_2, v_3\}$ and $_\mathscr{M}(T)_\mathscr{B} = \begin{pmatrix} 1 & 1 & 1 \\ 1 & \lambda & \mu \\ 1 & \lambda^2 & \mu^2 \end{pmatrix}$, find $_{\mathscr{M}'}(T)_\mathscr{B}$, where

$$\mathscr{M}' = \{u_1 + u_2 + u_3, u_2 + (\lambda + 1)u_3, u_3\}$$

Hence or otherwise, find the values of λ and μ for which the system of linear equations

$$x + y + z = 1$$
$$x + \lambda y + \mu z = 2$$
$$x + \lambda^2 y + \mu^2 z = 4$$

has (i) a unique solution (ii) more than one solution.

7. Find an **R**-basis for the vector space of all homogeneous real quadratic polynomials in three indeterminates x, y and z. Show that the mapping which takes such a polynomial $f(x, y, z)$ into $f(\alpha x + y + z, \beta y + \gamma z, 0)$ is a linear transformation and find its matrix relative to this **R**-basis.

8. Let U be the vector space of all real quadratic polynomials in two variables x and y and V the vector space of all real cubic polynomials in one variable x. Find **R**-bases for U and V. Define $T : U \to V$ by $T(f(x,y)) = f(x, 2)$. Show that T is a linear transformation and find the matrix of T relative to the **R**-bases for U and V.

4.4 The Kernel and Image of a Linear Transformation

Let V and W be K-spaces, and let T be a linear transformation of V into W. We now introduce certain subspaces of V and W which are important in later applications. Put

$$\ker T = \{v \in V \mid Tv = 0\}$$

and

$$\operatorname{im} T = \{T(v) \mid v \in V\}$$

Then the following lemma is proved.

LEMMA 4.7   (i) *ker T is a subspace of V,*
(ii) *im T is a subspace of W.*

PROOF   (i) ker T is non-empty since $0 \in \ker T$. If $v, v' \in \ker T$ and $\alpha \in K$, then $T(v) = T(v') = 0$ and $T(\alpha v + v') = \alpha T(v) + T(v') = 0$, that is $\alpha v + v' \in \ker T$ and ker T is a subspace of V.

(ii) im T is non-empty since $T(0) \in \operatorname{im} T$. If $w, w' \in \operatorname{im} T$ and $\alpha \in K$, then $w = T(v)$ and $w' = T(v')$ for some $v, v' \in V$, and thus

$$\alpha w + w' = \alpha T(v) + T(v')$$
$$= T(\alpha v + v')$$

where $\alpha v + v' \in V$, i.e. $\alpha w + w' \in \operatorname{im} T$ and im T is a subspace of W. ∎

If V and W are finite dimensional vector spaces over K, then ker T and im T are also finite dimensional over K and we can give the following

DEFINITION 4.8   *nullity $T = (\ker T : K)$, rank $T = (\operatorname{im} T : K)$.*

The rank and nullity of a linear transformation are connected as follows.

THEOREM 4.9   *If V and W are finite dimensional vector spaces over K and T a linear transformation of V into W then*

$$\textit{nullity } T + \textit{rank } T = (V : K)$$

PROOF   Suppose that nullity $T = r$ and let $\{v_1, v_2, \ldots, v_r\}$ be a K-basis for ker T. By Theorem 3.17, this K-basis for ker T can be extended to give a K-basis $\{v_1, \ldots, v_r, v_{r+1}, \ldots, v_n\}$ for V, if $n = (V:K)$. By definition of im T, it is clear that $\{Tv_1, Tv_2, \ldots, Tv_n\}$ generates im T over K. But $Tv_1 = Tv_2 = \ldots = Tv_r = 0$, since $v_1, v_2, \ldots, v_r \in \ker T$, which implies that $\{Tv_{r+1}, \ldots, Tv_n\}$ generates im T over K. We show that this set is also linearly independent over K and hence is a K-basis for im T and the result will follow, i.e. $r + (n-r) = n$.

Consider

$$\alpha_{r+1} T(v_{r+1}) + \ldots + \alpha_n T(v_n) = 0$$

then

$$T(\alpha_{r+1} v_{r+1} + \ldots + \alpha_n v_n) = 0$$

and $\alpha_{r+1} v_{r+1} + \ldots + \alpha_n v_n \in \ker T$. But $\{v_1, v_2, \ldots, v_r\}$ is a K-basis for $\ker T$ and thus

$$\alpha_{r+1} v_{r+1} + \ldots + \alpha_n v_n = \beta_1 v_1 + \ldots + \beta_r v_r$$

for $\beta_1, \beta_2, \ldots, \beta_r \in K$, or in other words

$$\beta_1 v_1 + \ldots + \beta_r v_r - \alpha_{r+1} v_{r+1} - \ldots - \alpha_n v_n = 0$$

Since $\{v_1, \ldots, v_n\}$ is a K-basis for V and is linearly independent over K, we have in particular that $\alpha_{r+1} = \ldots = \alpha_n = 0$ and $\{T(v_{r+1}), \ldots, T(v_n)\}$ is linearly independent over K. ∎

EXAMPLES

1. Let $T: V_3(\mathbf{R}) \to V_3(\mathbf{R})$ be defined by

$$T(\alpha_1, \alpha_2, \alpha_3) = (\alpha_1 + 2\alpha_2 - \alpha_3, 2\alpha_1 + \alpha_3, \alpha_1 - 2\alpha_2 + 2\alpha_3)$$

Then T is a linear transformation on $V_3(\mathbf{R})$. We now determine $\mathbf{R}$-bases for $\ker T$ and im T. We have that

$$\begin{aligned} \text{im } T = \{ T(\alpha_1, \alpha_2, \alpha_3) &= \alpha_1(1, 2, 1) + \alpha_2(2, 0, -2) \\ &+ \alpha_3(-1, 1, 2) \,|\, \alpha_1, \alpha_2, \alpha_3 \in \mathbf{R} \} \end{aligned}$$

Thus $\{(1, 2, 1), (2, 0, -2), (-1, 1, 2)\}$ generates im T over $\mathbf{R}$. Since

$$(-1, 1, 2) = \tfrac{1}{2}(1, 2, 1) - \tfrac{3}{4}(2, 0, -2)$$

and $\{(1, 2, 1), (2, 0, -2)\}$ is clearly linearly independent over $\mathbf{R}$ it follows that $\{(1, 2, 1), (2, 0, -2)\}$ is an $\mathbf{R}$-basis for im T and rank $T = 2$. By the above theorem, we have that nullity $T = (V_3(\mathbf{R}):\mathbf{R}) - 2 = 1$. Now, $(\alpha_1, \alpha_2, \alpha_3) \in \ker T$ if and only if

107

$$\alpha_1 + 2\alpha_2 - \alpha_3 = 0$$

$$2\alpha_1 \qquad + \alpha_3 = 0$$

$$\alpha_1 - 2\alpha_2 + 2\alpha_3 = 0$$

From

$$\begin{pmatrix} 1 & 2 & -1 \\ 2 & 0 & 1 \\ 1 & -2 & 2 \end{pmatrix} \rightarrow \begin{pmatrix} 1 & 2 & -1 \\ 0 & -4 & 3 \\ 0 & -4 & 3 \end{pmatrix} \rightarrow \begin{pmatrix} 1 & 0 & \frac{1}{2} \\ 0 & 1 & -\frac{3}{4} \\ 0 & 0 & 0 \end{pmatrix}$$

we deduce that this system reduces to

$$\alpha_1 = -\tfrac{1}{2}\alpha_3, \quad \alpha_2 = \tfrac{3}{4}\alpha_3$$

or, in other words

$$\ker T = \{\alpha_3(-\tfrac{1}{2}, \tfrac{3}{4}, 1) \mid \alpha_3 \in \mathbf{R}\}$$

and $\{(-2, 3, 4)\}$ is an $\mathbf{R}$-basis for $\ker T$.

2. Let $T : C^2(\mathbf{R}) \rightarrow C(\mathbf{R})$ be defined by

$$T(y) = \frac{d^2 y}{dx^2} + y, \qquad y \in C^2(\mathbf{R})$$

Then it is easily verified that T is a linear transformation.

Now

$$\ker T = \{ y \in C^2(\mathbf{R}) \mid T(y) = 0 \}$$

$$= \{ y \in C^2(\mathbf{R}) \mid \frac{d^2 y}{dx^2} + y = 0 \}$$

Thus, $\ker T$ is the solution space of the differential equation

$$\frac{d^2 y}{dx^2} + y = 0$$

which was completely determined in Example 8, p. 81.

3. Let $T : V_4(\mathbf{R}) \rightarrow V_2(\mathbf{R})$ be defined by

$$T(x, y, z, w) = AX$$

where $A = \begin{pmatrix} 1 & -2 & 3 & -1 \\ 2 & 1 & -1 & 1 \end{pmatrix}$, $X^t = (x, y, z, w)$, then

$$\ker T = \{(x, y, z, w) \mid AX = 0\}$$

108

Thus, ker T is the solution space of the system of linear equations determined in Example 7, p. 80.

The application of the ideas introduced in this section and illustrated by this last example to the solution of linear equations are developed in the final section of this chapter. For a discussion on the corresponding problem for linear differential equations see D.L. Kneider, R.G. Kuller, D.R. Ostberg and F.W. Perkins (*loc. cit.*).

Exercises 4.4

1. Let $T: V_4(\mathbf{R}) \to V_3(\mathbf{R})$ be the linear transformation defined by

$$T(\alpha_1, \alpha_2, \alpha_3, \alpha_4) = (\alpha_1 - \alpha_2 + \alpha_3 + \alpha_4, \alpha_1 + 2\alpha_2 - \alpha_3 + \alpha_4,$$
$$3\alpha_2 - 2\alpha_3)$$

Find $\mathbf{R}$-bases for ker T and im T.

2. Find the rank and nullity of the linear transformation from $V_4(\mathbf{R})$ into $V_3(\mathbf{R})$ whose matrix relative to standard bases is

$$\begin{pmatrix} 1 & 2 & -1 & 2 \\ 2 & 6 & 3 & -3 \\ 0 & 2 & 5 & -7 \end{pmatrix}$$

3. Find ker T and im T for all the linear transformations defined in Exercises 4.1, No. 1.

4. Find the rank and nullity of a linear transformation T from $V_4(\mathbf{R})$ into $V_3(\mathbf{R})$ defined by

$$T(\alpha_1, \alpha_2, \alpha_3, \alpha_4) = (\alpha_1 - \alpha_3 + 2\alpha_4, -2\alpha_1 + \alpha_2 + 2\alpha_3, \alpha_2 + 4\alpha_4)$$

Show that $(1, 3, k)$ is in im T if and only if $k = 5$.
Find the condition for $(1, x, 1, y)$ to be in ker T.

5. Let V denote the real vector space of polynomials $f(x, y)$ with real coefficients of degree not exceeding n in two variables x and y. Show that the mappings S and T defined by

$$S(f(x, y)) = x\frac{\partial f}{\partial x} + y\frac{\partial f}{\partial y}, T(f(x, y)) = x^2\frac{\partial^2 f}{\partial x^2} + y^2\frac{\partial^2 f}{\partial y^2}$$

are linear transformations on V. Find the kernel and image of S and T. Find an $\mathbf{R}$-basis for V and find the matrices of S and T relative to this $\mathbf{R}$-basis.

6. Let V be the vector space of real functions which have derivatives of all orders. If D is the derivative, find

(i) $\ker D$,   (ii) $\ker D^n (n \geqslant 1)$ and   (iii) $\ker(D - I)$.

7. If $S = \begin{pmatrix} 1 & 1 \\ 1 & 2 \end{pmatrix}$, find **R**-bases for $\ker T$ and $\operatorname{im} T$ for all the linear transformations defined in Exercises 4.2, No. 4.

8. If T is a linear transformation of a vector space V into a vector space W, show that the elements of V which are mapped into a given subspace U of W, form a subspace X of V. If the dimensions of V, W, U, X are m, n, p, q respectively and if the rank of T is n, find a relation between m, n, p, q.

9. Let V be an n-dimensional K-space and S and T are linear transformations on V, prove that

$$\text{nullity } (ST) \leqslant \text{nullity } S + \text{nullity } T$$

If $S^n = 0$, but $S^{n-1} \neq 0$, determine nullity S.

10. If $S : U \to V$ and $T : V \to U$ are linear transformations, prove that

$$\text{rank } T - \text{rank } ST \leqslant (U : K) - \text{rank } S$$

11. Find a linear transformation T on some vector space V such that $\ker T = \operatorname{im} T$. Can this be done for all vector spaces?

12. If V is a K-space, prove that
$\operatorname{im} T \cap \ker T = \{0\}$ if and only if $T(Tv) = 0$ implies $Tv = 0$, where $v \in V$.

13. If T is a linear transformation on a finite dimensional K-space V and rank $T^2 = \text{rank } T$, then $\operatorname{im} T \cap \ker T = \{0\}$.

14. If T is a linear transformation on V such that $T^2 = T$, prove that

   (i) $\ker T = \operatorname{im}(I - T)$, $\ker(I - T) = \operatorname{im} T$
   (ii) $\ker T \cap \operatorname{im} T = 0$
  (iii) every $v \in V$ can be uniquely expressed in the form $v = v_1 + v_2$, where $v_1 \in \ker T$, $v_2 \in \operatorname{im} T$.

4.5 *K*-ismorphisms and Non-singular Linear Transformations

We now connect these ideas with other important concepts in algebra, namely K-isomorphisms and non-singular or invertible linear transformations.

DEFINITION 4.10   *A linear transformation (K-homomorphism)*
$T : V \to W$ *is called*

(i) *a **K-isomorphism** if T is a bijective mapping, i.e. T is injective*
$(T(v) = T(v')$ *implies* $v = v')$ *and T is surjective (if* $w \in W$, *there exists*
a $v \in V$ *such that* $Tv = w$).

(ii) *a **non-singular transformation** if ker* $T = \{0\}$.

Let $T : V \to W$ and $S : W \to U$ be K-isomorphisms, where V, W, U are
K-spaces, then $ST : V \to U$ defined by $(ST)(v) = S(Tv)$ for all $v \in V$ is
bijective, and is also a linear transformation since if $\alpha \in K$, $v, v' \in V$
then

$$
\begin{aligned}
(ST)(\alpha v + v') &= S(T(\alpha v + v')) \\
&= S(\alpha T(v) + T(v')) \\
&= \alpha S(T(v)) + S(T(v')) \\
&= \alpha ST(v) + ST(v')
\end{aligned}
$$

If T is a K-isomorphism of V onto W, then since T is a bijective
mapping $T^{-1} : W \to V$ is also a bijective mapping and is a linear
transformation from W onto V since if $w_1, w_2 \in W$ and $\alpha \in K$, then
$w_1 = Tv_1, w_2 = Tv_2$ for unique $v_1, v_2 \in V$ and

$$
\begin{aligned}
T^{-1}(\alpha w_1 + w_2) &= T^{-1}(\alpha Tv_1 + Tv_2) \\
&= T^{-1}T(\alpha v_1 + v_2) \\
&= \alpha v_1 + v_2 \\
&= \alpha T^{-1}w_1 + T^{-1}w_2
\end{aligned}
$$

Thus, T is an **invertible** linear transformation. Conversely, if T is an
invertible linear transformation, then T is a K-isomorphism and thus
these two concepts are equivalent.

We now show that if the two K-spaces V and W are finite dimensional
and have the same dimension then the concepts of invertible and non-
singular linear transformations are equivalent. In fact, we show that
these are equivalent to various other statements.

THEOREM 4.11   *Let V and W be finite dimensional K-spaces with*
$(V : K) = (W : K) = n$ *and T a linear transformation of V into W. Then,*
the following statements are equivalent

(i) *T is a K-isomorphism*
(ii) *T is invertible*
(iii) *T is injective*

(iv) *T is non-singular*

(v) *rank T = n*

(vi) *T is surjective*

(vii) *if $\{v_1, v_2, \ldots, v_n\}$ is a K-basis for V, then $\{T(v_1), T(v_2), \ldots, T(v_n)\}$ is a K-basis for W.*

PROOF   In the proof we show that

(i) ⇒ (ii) ⇒ (iii) ⇒ (iv) ⇒ (v) ⇒ (vi) ⇒ (vii) ⇒ (i),

which will imply that all seven statements are equivalent.

(i) ⇒ (ii) has been proved above.

(ii) ⇒ (iii) is a well-known for maps in general.

(iii) ⇒ (iv). Let $v \in \ker T$, then $T(v) = 0 = T(0)$ and as T is injective $v = 0$ and $\ker T = \{0\}$ and T is non-singular.

(iv) ⇒ (v). If T is non-singular $\ker T = \{0\}$ and nullity $T = 0$. By Theorem 4.9 we have rank $T = (V : K) = n$.

(v) ⇒ (vi). By Lemma 4.7, im T is a subspace of W. If rank $T = n = (W : K)$ then im $T = W$ and T is surjective.

(vi) ⇒ (vii). If T is surjective, then if $w \in W$, there exists a $v \in V$ such that $Tv = w$. If $\{v_1, v_2, \ldots, v_n\}$ is a K-basis for V then

$$v = \sum_{i=1}^{n} \alpha_i v_i, \alpha_i \in K \ (i = 1, \ldots, n) \text{ and } w = Tv = \sum_{i=1}^{n} \alpha_i T(v_i),$$

i.e. $\{T(v_1), \ldots, T(v_n)\}$ generates W. By Corollary 1 to Theorem 3.17 this must be a K-basis for W.

(vii) ⇒ (i). Let $v, v' \in V$ such that $T(v) = T(v')$. We have

$$v = \sum_{i=1}^{n} \alpha_i v_i, v' = \sum_{i=1}^{n} \beta_i v_i, \alpha_i, \beta_i \in K \ (i = 1, \ldots, n) \text{ and thus}$$

$$\sum_{i=1}^{n} (\alpha_i - \beta_i) T(v_i) = 0.$$

But $\{T(v_1), \ldots, T(v_n)\}$ is a K-basis for W and so $\alpha_i = \beta_i \ (i = 1, \ldots, n)$,

i.e. $v = v'$ and T is injective. Let $w \in W$, then $w = \sum_{i=1}^{n} \alpha_i T(v_i)$, or

$$T\left(\sum_{i=1}^{n} \alpha_i v_i\right) = w \text{ with } \sum_{i=1}^{n} \alpha_i v_i \in V \text{ and } T \text{ is surjective. Thus } T \text{ is a}$$

K-isomorphism. ∎

In particular, we note that if T is a linear transformation on a finite dimensional vector space V then the concepts of K-isomorphism, invertible and non-singular are equivalent, and furthermore, to prove that a linear transformation is invertible we need only verify that it is either injective or surjective.

112

We now prove two useful results concerning non-singular linear transformations.

THEOREM 4.12   *If V is a finite dimensional K-space and S, T are linear transformations on V where T is non-singular, then*

$$rank\ (TS)\ =\ rank\ S\ =\ rank\ (ST)$$

PROOF   If $v \in im(ST)$ then $v = (ST)(v') = S(T(v'))$ for some $v' \in V$, i.e. $v \in im\ S$ and $im(ST) \subseteq im\ S$. Conversely, if $v \in im\ S$, then $v = S(v')$ for some $v' \in V$. But T is non-singular and so in particular is surjective and so there exists $v'' \in V$ such that $T(v'') = v'$, i.e. $v = ST(v'')$, $v \in im(ST)$ and $im\ S = im\ ST$ and rank S = rank ST.

Now if $v \in ker\ S$ then $S(v) = 0$ and so $TS(v) = 0$, i.e. $v \in ker\ TS$ and $ker\ S \subseteq ker\ TS$. If $v \in ker\ TS$ then $(TS)(v) = 0$, but T is non-singular and so $S(v) = 0$ and $v \in ker\ S$, i.e. $ker\ S = ker\ TS$.

Hence, we have nullity (TS) = nullity (S) and by Theorem 4.9, it follows that rank TS = rank S as required.   █

THEOREM 4.13   *If S and T are linear transformations on a finite dimensional K-space then ST is non-singular if and only if S and T are non-singular. If ST is non-singular, then $(ST)^{-1} = T^{-1}S^{-1}$.*

PROOF   If ST is non-singular then $ker\ ST = \{0\}$.
But $ker\ T \subseteq ker\ ST$, thus $ker\ T = \{0\}$ and T is non-singular. By Theorem 4.12 it follows that

$$rank\ (ST)\ =\ rank\ S\ =\ (V:K)$$

and by Theorem 4.11 S is also non-singular.

If S and T are non-singular then by Theorem 4.12

$$rank\ (ST)\ =\ rank\ S\ =\ (V:K)$$

and by Theorem 4.11, ST is non-singular.

If ST is non-singular, then ST has a unique inverse $(ST)^{-1}$, i.e. $(ST)(ST)^{-1} = I_V$. But, in addition, we have

$$(ST)(T^{-1}S^{-1})\ =\ S(TT^{-1})S^{-1}\ =\ I_V$$

and thus $(ST)^{-1} = T^{-1}S^{-1}$.   █

The concept of K-isomorphism is of sufficient importance to deserve further attention.

If $T: V \rightarrow W$ is a K-isomorphism, then we have seen above that $T^{-1}: W \rightarrow V$ is also a K-isomorphism. Indeed K-isomorphy is an equivalence relation on the set of all K-spaces. Clearly $I_V: V \rightarrow V$ is a

K-isomorphism and if $T : V \to W$ and $S : W \to U$ are K-isomorphisms then $ST : V \to U$ is also a K-isomorphism. In this case, we say that V and W are **K-isomorphic** and we denote this by $V \cong W$. K-spaces which are K-isomorphic are regarded as being "equal" or "the same", although they may contain different elements and the operations of addition and scalar multiplication are distinct. The reader may have noticed for example the similarity of the three examples (i) $V_4(\mathbf{R})$, (ii) $P_3(\mathbf{R})$, the $\mathbf{R}$-space of real polynomials of degree $\leqslant 3$, (iii) $M_2(\mathbf{R})$, and that in practice they are dealt with in the same way. The isomorphism is easily established, for

$$T : P_3(\mathbf{R}) \to V_4(\mathbf{R}) \quad \text{given by}$$

$$T(\alpha_0 + \alpha_1 x + \alpha_2 x^2 + \alpha_3 x^3) = (\alpha_0, \alpha_1, \alpha_2, \alpha_3)$$

is a $\mathbf{R}$-isomorphism and

$$S : M_2(\mathbf{R}) \to V_4(\mathbf{R}) \quad \text{given by}$$

$$S \begin{pmatrix} \alpha & \beta \\ \gamma & \delta \end{pmatrix} = (\alpha, \beta, \gamma, \delta)$$

is also a $\mathbf{R}$-isomorphism. We shall not deal with these examples in detail because the following more general theorem can be proved.

THEOREM 4.14   *Let V be a finite dimensional K-space of dimension n, then $V \cong V_n(K)$.*

PROOF   If $(V : K) = n$, let $\{v_1, v_2, \ldots, v_n\}$ be a K-basis for V. If $v \in V$ then $v = \alpha_1 v_1 + \ldots + \alpha_n v_n$, where $\alpha_i \in K \, (i = 1, \ldots, n)$ are uniquely determined. Then $T : V \to V_n(K)$ defined by

$$T(v) = T(\alpha_1 v_1 + \ldots + \alpha_n v_n) = (\alpha_1, \alpha_2, \ldots, \alpha_n)$$

is a well-defined map which is easily shown to be a linear transformation over K. T is clearly a surjective map and since $(V : K) = (V_n(K) : K) = n$, it follows from Theorem 4.11 that T is a K-isomorphism. ∎

This theorem is a strong result — if our goal had been to classify all finite dimensional vector spaces over a field K then it tells us that every finite dimensional vector space over K is essentially a $V_n(K)$ for some positive integer n. This suggests that rather than deal with an "abstract" K-space it is only necessary to consider the "more concrete" $V_n(K)$. However, in practice, it turns out that there are sometimes advantages in working at the more abstract level — for example, simple and elegant proofs may be possible which may not be immediately apparent at the more concrete level. The detailed and sometimes cumbersome

114

explicit information available when more concrete vector spaces comprising vectors, matrices or polynomials are considered may blind us from appreciating what the bare essentials to carry through a certain proof may be. But, as is typical in mathematics, developments are usually made by exploiting the interplay between the abstract and concrete — when no more progress is possible at the abstract level, something may be done by considering an isomorphic concrete example or vice versa.

It could be argued that a more elegant and natural introduction to linear algebra would be to first consider abstract vector spaces and their linear transformations — develop their theory as far as possible and then introduce matrices and the applications to linear equations. In this way, some of the more cumbersome and tedious proofs could be eliminated and the advantages of working at the abstract level would become more apparent. However, the philosophy of our approach is that there are more benefits to be gained, especially to newcomers to the subject, by first working at the more concrete and familiar level and use this as a firm foundation from which the new abstract concepts can be introduced. In any case, for explicit computations the work in the earlier chapters will necessarily eventually have to be covered.

The same is also true relative to linear transformations and matrices. We have seen in §4.2, that every linear transformation can be represented by a matrix, but whereas the proof of the associativity of multiplication of linear transformations is trivial, the corresponding result for multiplication of matrices, although elementary, is cumbersome.

A further important isomorphism of K-spaces is given as follows:

Let $\mathscr{L}(V, W)$ denote the set of all linear transformations of a K-space V into a K-space W. If $S, T \in \mathscr{L}(V, W)$ their sum $S + T$ is defined by

$$(S + T)(v) = S(v) + T(v) \quad \text{for all} \quad v \in V$$

If $v, v' \in V, \alpha \in K$, then

$$(S + T)(\alpha v + v') = S(\alpha v + v') + T(\alpha v + v')$$
$$= \alpha S(v) + S(v') + \alpha T(v) + T(v')$$
$$= \alpha (S + T)(v) + (S + T)(v')$$

i.e. $S + T \in \mathscr{L}(V, W)$.

Similarly, if $\alpha \in K, S \in \mathscr{L}(V, W)$, define

$$(\alpha S)(v) = \alpha S(v) \quad \text{for all} \quad v \in V$$

then $\alpha S \in \mathscr{L}(V, W)$.

With these definitions of addition and scalar multiplication it can be proved that

LEMMA 4.15 $\mathscr{L}(V, W)$ *is a K-space.*

PROOF Exercise.

If V and W are finite dimensional vector spaces, then the following important K-isomorphism can be established.

THEOREM 4.16 *If V and W are finite dimensional K-spaces, where* $(V:K) = n$, $(W:K) = m$ *then*

$$\mathscr{L}(V, W) \cong M_{m,n}(K)$$

PROOF If $T \in \mathscr{L}(V, W)$ and $\mathscr{B} = \{v_1, \ldots, v_n\}$, $\mathscr{B}' = \{w_1, \ldots, w_m\}$ are K-bases for V, W respectively, let $_{\mathscr{B}}(T)_{\mathscr{B}'} = (\alpha_{ij})$ be the matrix of T relative to $\mathscr{B}$ and $\mathscr{B}'$. Define $\phi : \mathscr{L}(V, W) \to M_{m,n}(K)$ by

$$\phi(T) = {}_{\mathscr{B}}(T)_{\mathscr{B}'}$$

If $\alpha \in K$, $S, T \in \mathscr{L}(V, W)$ and $_{\mathscr{B}}(S)_{\mathscr{B}'} = (\beta_{ij})$ then

$$\phi(\alpha T + S) = (\alpha\alpha_{ij} + \beta_{ij}) = \alpha\phi(T) + \phi(S)$$

since

$$(\alpha T + S)v_j = \sum_{i=1}^{m} (\alpha\alpha_{ij} + \beta_{ij})w_i \qquad (j = 1, \ldots, n)$$

i.e. ϕ is a linear transformation.

ϕ is surjective, since by §4.2 we see that if $A \in M_{m,n}(K)$ then there exists a $T_A \in \mathscr{L}(V, W)$ such that $\phi(T_A) = A$. Furthermore, if $S, T \in \mathscr{L}(V, W)$ and $\phi(S) = \phi(T)$, i.e. $(\alpha_{ij}) = (\beta_{ij})$ then $\alpha_{ij} = \beta_{ij}$ $(i = 1, \ldots, m; j = 1, \ldots, n)$ and thus for $j = 1, \ldots, n$

$$Tv_j = \sum_{i=1}^{m} \alpha_{ij} w_i = \sum_{i=1}^{m} \beta_{ij} w_i = Sv_j$$

and so $S = T$, i.e. ϕ is injective and we have the required K-isomorphism. ∎

COROLLARY *If V and W are finite dimensional K-spaces where* $(V:K) = n$, $(W:K) = m$ *then* $(\mathscr{L}(V, W):K) = mn$.

We have seen that in both $M_n(K)$ and $\mathscr{L}(V, V) = \mathscr{L}(V)$ in addition to the two operations of addition and scalar multiplication, the operation of "multiplication" is also possible. It is of interest to see whether this operation is also preserved under the mapping ϕ; to be more precise, if $S, T \in \mathscr{L}(V)$, is $\phi(ST) = \phi(S)\phi(T)$ or

$(ST)_{\mathscr{B}} = (S)_{\mathscr{B}} (T)_{\mathscr{B}}$? This is easily verified to be the case for if $\mathscr{B} = \{v_1, \ldots, v_n\}$ and

$$Tv_j = \sum_{i=1}^{n} \alpha_{ij} v_i \qquad (j = 1, \ldots, n)$$

$$Sv_j = \sum_{i=1}^{n} \beta_{ij} v_i \qquad (j = 1, \ldots, n)$$

Then $(ST)(v_j) = S\left(\sum_{i=1}^{n} \alpha_{ij} v_i \right)$

$$= \sum_{i=1}^{n} \alpha_{ij} \left(\sum_{k=1}^{n} \beta_{ki} v_k \right)$$

$$= \sum_{k=1}^{n} \left(\sum_{i=1}^{n} \beta_{ki} \alpha_{ij} \right) v_k$$

and the (k, j)-element of the matrix $(ST)_{\mathscr{B}}$ is $\sum_{i=1}^{n} \beta_{ki} \alpha_{ij}$, which is the (k, j)-element of the matrix $(S)_{\mathscr{B}} (T)_{\mathscr{B}}$.

The reader may have wondered when the matrix of a linear transformation was first introduced why one did not simply define the more natural

$$Tv_i = \sum_{j=1}^{n} \alpha_{ij} v_j \qquad (i = 1, \ldots, n)$$

where the ordering of the i, j is preserved, i.e. the matrix is the transpose of the one defined in §4.2. If this had been done, then we would now have obtained $(ST)_{\mathscr{B}} = (T)_{\mathscr{B}} (S)_{\mathscr{B}}$ or $\phi(ST) = \phi(T) \phi(S)$. (The ideas developed above can be expounded more succinctly as follows: A set $\mathscr{A}$ together with operations of addition, multiplication and scalar multiplication by elements of a field K is called a **linear K-algebra** if

(i) $\mathscr{A}$ is a K-space

(ii) if $a, b \in \mathscr{A}$ then $ab \in \mathscr{A}$

(iii) $(ab)c = a(bc), a(b + c) = ab + ac, (a + b)c = ac + bc$ for all $a, b, c \in \mathscr{A}$.

(iv) $\alpha(ab) = (\alpha a) b = a(\alpha b)$ for all $a, b \in \mathscr{A}, \alpha \in K$.

If $\mathscr{A}$ and $\mathscr{A}'$ are linear K-algebras then $\mathscr{A}$ and $\mathscr{A}'$ are **algebra-isomorphic** if there exists a bijective mapping $\phi : \mathscr{A} \to \mathscr{A}'$ such that

117

(i) $\phi(\alpha a + a') = \alpha\phi(a) + \phi(a')$
   for all $\alpha \in K, a, a' \in \mathscr{A}$

(ii) $\phi(ab) = \phi(a)\,\phi(b)$
   for all $a, b \in \mathscr{A}$.

Then $M_n(K)$ and $\mathscr{L}(V)$ are linear K-algebras and if $(V:K) = n$, then $M_n(K)$ and $\mathscr{L}(V)$ are not only K-isomorphic but algebra-isomorphic.)

Exercises 4.5

1. If $\{v_1, v_2, v_3, v_4\}$ is an **R**-basis for the vector space V, for what value of λ is the linear transformation T defined by

$$Tv_1 = v_1 + \lambda v_4$$

$$Tv_i = 2v_{i-1} + v_i \qquad (i = 2, 3, 4)$$

non-singular?

2. If T is the linear transformation on $V_3(\mathbf{R})$ defined by

$$T(\alpha_1, \alpha_2, \alpha_3) = (3\alpha_1 - \alpha_2, \alpha_1 - \alpha_2 + \alpha_3, -\alpha_1 + 2\alpha_2 - \alpha_3)$$

show that T is non-singular. Give a rule for T^{-1} like the one which defines T.

3. A linear transformation on **C** regarded as **R**-space is defined by $T(z) = (1 - i)z$ for all $z \in \mathbf{C}$, show that T is non-signular.

4. Prove that the differentiation transformation D on $P_n(\mathbf{R})$ is singular. What does this imply for the kernel of D?

5. If T is a linear transformation on V such that $T^2 = 0$, show that $I - T$ is non-singular.

4.6 Applications to Linear Equations and the Rank of Matrices

Let $A = (\alpha_{ij}) \in M_{m,n}(K)$ and let $c_j (j = 1, \ldots, n)$ be the n columns of A, i.e. $c_j = (\alpha_{1j}, \alpha_{2j}, \ldots, \alpha_{mj})$. Then $\{c_1, c_2, \ldots, c_n\}$ generates a subspace C_A of $V_m(K)$.

DEFINITION 4.17 *If $A \in M_{m,n}(K)$ the **column rank** of A is defined to be $(C_A:K)$, or in other words, it is the maximum number of linearly independent column vectors in A.*

EXAMPLES

1. Let

$$A = \begin{pmatrix} 1 & -1 & 2 & 1 \\ 2 & 0 & 1 & 1 \\ 3 & -1 & 3 & 2 \\ 0 & 2 & -3 & -1 \end{pmatrix}$$

If c_1, c_2, c_3 and c_4 are the columns of A and we consider

$$\alpha_1 c_1 + \alpha_2 c_2 + \alpha_3 c_3 + \alpha_4 c_4 = 0$$

then we have a system of four linear homogeneous equations in the variables α_1, α_2, α_3, α_4 with matrix of coefficients A. It is easily shown that A is row equivalent to

$$\begin{pmatrix} 1 & 0 & \frac{1}{2} & \frac{1}{2} \\ 0 & 1 & -\frac{3}{2} & -\frac{1}{2} \\ 0 & 0 & 0 & 0 \\ 0 & 0 & 0 & 0 \end{pmatrix}$$

and thus

$$\alpha_1 = -\tfrac{1}{2}\alpha_3 - \tfrac{1}{2}\alpha_4$$
$$\alpha_2 = \tfrac{3}{2}\alpha_3 + \tfrac{1}{2}\alpha_4$$

from which, by putting $\alpha_3 = 1$, $\alpha_4 = 0$ and $\alpha_3 = 0$, $\alpha_4 = 1$, we obtain

$$c_3 = \tfrac{1}{2}(c_1 - 3c_2), \quad c_4 = \tfrac{1}{2}(c_1 - c_2)$$

respectively. $\{c_1, c_2\}$ is linearly independent over $\mathbf{R}$ and the column rank of A is 2.

2. Find the column rank of the matrix

$$A = \begin{pmatrix} 2a+b & 3a+3b & -a-2b & 2a+b \\ 3b & 3b & -a+b & a+5b \\ 3a & 4a+2b & -3b & 3a \\ b-a & -2a-b & 3b & 3b \end{pmatrix}$$

for all values of a and b.

From Theorem 3.13 we see that the columns of A are linearly independent if and only if $\det A \neq 0$.

119

Now

$$\det A = \begin{vmatrix} -a-2b & 3a+3b & -a-2b & 2a+b \\ 0 & 3b & -a+b & a+5b \\ -a-2b & 4a+2b & -3b & 3a \\ a+2b & -2a-b & 3b & 3b \end{vmatrix}$$

$$= (a+2b) \begin{vmatrix} 0 & a+2b & -a+b & 2a+4b \\ 0 & 3b & -a+b & a+5b \\ 0 & 2a+b & 0 & 3a+3b \\ 1 & -2a-b & 3b & 3b \end{vmatrix}$$

$$= -(a+2b)(b-a) \begin{vmatrix} a-b & 0 & a-b \\ 3b & 1 & a+5b \\ 2a+b & 0 & 3a+3b \end{vmatrix}$$

$$= (a+2b)(a-b)^2 \begin{vmatrix} 1 & 0 \\ 2a+b & a+2b \end{vmatrix}$$

$$= (a+2b)^2 (a-b)^2$$

Thus, column rank $A = 4$ unless $a = b$ or $a = -2b$.
If $a = b = 0$, then clearly column rank $A = 0$.
If $a = b \neq 0$, then

$$A = \begin{pmatrix} 3a & 6a & -3a & 3a \\ 3a & 3a & 0 & 6a \\ 3a & 6a & -3a & 3a \\ 0 & -3a & 3a & 3a \end{pmatrix}$$

We note that $c_3 = c_1 - c_2$ and $c_4 = 3c_1 - c_2$ and if $a \neq 0$ $\{c_1, c_2\}$ is linearly independent over $\mathbf{R}$, that is, if $a = b \neq 0$, column rank $A = 2$.
If $a = -2b \neq 0$, then

$$A = \begin{pmatrix} -3b & -3b & 0 & -3b \\ 3b & 3b & 3b & 3b \\ -6b & -6b & -3b & -6b \\ 3b & 3b & 3b & 3b \end{pmatrix}$$

120

and $c_1 = c_2 = c_4$ and $\{c_1, c_3\}$ is linearly independent over **R**, that is, column rank $A = 2$.

Before proceeding, we note that the **row rank** of A may be defined to be the maximum number of linearly independent row vectors in A. We show below that the row rank of a matrix is equal to the column rank of A. In the above example 1, note that if the rows are denoted by r_1, r_2, r_3, r_4 respectively, then $r_3 = r_1 + r_2, r_4 = r_2 - 2r_1, \{r_1, r_2\}$ is linearly independent over **R** and so the row rank of $A = 2$.

If $A = (\alpha_{ij}) \in M_{m,n}(K)$, let $T : V_n(K) \to V_m(K)$ be the linear transformation defined by

$$T(x_1, \ldots, x_n) = (y_1, \ldots, y_m)$$

where $y_i = \sum\limits_{j=1}^{n} \alpha_{ij} x_j$ $(i = 1, \ldots, m)$. Then A is the matrix of T relative to the standard K-bases for $V_n(K)$ and $V_m(K)$.

In the previous section, we defined the rank T, we can now prove the reassuring theorem that

THEOREM 4.18   *rank $T =$ column rank A.*

PROOF   By definition, rank $T = (\operatorname{im} T : K)$.
If

$$(y_1, \ldots, y_m) \in \operatorname{im} T, \text{ then } y_i = \sum\limits_{j=1}^{n} \alpha_{ij} x_j \qquad (i = 1, \ldots, m)$$

or

$$(y_1, \ldots, y_m) = \sum\limits_{j=1}^{n} x_j (\alpha_{1j}, \alpha_{2j}, \ldots, \alpha_{mj})$$

Thus $(y_1, \ldots, y_m)$ is a linear combination of the columns of A and the columns of A generate im T over K. Hence rank $T =$ column rank of A as required. ∎

As a corollary to this theorem we can now prove some important statements concerning the column rank of a matrix.

COROLLARY   (i) *A matrix $A \in M_n(K)$ is invertible if and only if column rank $A = n$.*

(ii) *If $A \in M_n(K)$ is invertible, then the column ranks of AB, BA and B are equal, where B is any n-rowed matrix.*

The proof of (ii) uses Theorem 4.12.
Furthermore we can prove

THEOREM 4.19   *Row equivalent matrices have the same column rank.*

121

PROOF   If B is row equivalent to A, then by Theorem 1.17 there exist elementary matrices $E_1, E_2, \ldots, E_k$ such that

$$B = E_1 E_2 \ldots E_k A$$

But elementary matrices are invertible and so by (ii) above

column rank of B = column rank of A ∎

We can now prove that the row and column ranks of a matrix are equal. We first note

THEOREM 4.20   *If a matrix B is obtained from a matrix A by a single elementary row operation then*

   row rank B = row rank A

PROOF   This result follows immediately on consideration of the three types of elementary row operations separately. ∎

COROLLARY   *Row equivalent matrices have the same row rank. Equivalent matrices have the same row rank.*

From this corollary and Theorem 4.19 we now obtain

THEOREM 4.21   *row rank A = column rank A.*

PROOF   Let R be the reduced echelon matrix of A. If R has r non-zero rows, then consideration of the form of R implies that
row rank R = column rank $R = r$. Then, by the above corollary and Theorem 4.19, we have row rank A = column rank $A = r$. ∎

From now on we refer to the **rank** of a matrix only.

Now consider the system of m linear equations in the n variables $x_1, x_2, \ldots, x_n$,

$$\sum_{j=1}^{n} \alpha_{ij} x_j = \beta_i \qquad (i = 1, 2, \ldots, m)$$

Then, by the above, we see that this represents the linear transformation $T : V_n(K) \to V_m(K)$ defined by

$$T(x_1, \ldots, x_n) = (\beta_1, \beta_2, \ldots, \beta_m)$$

We prove two theorems concerning the solution of linear equations. We first consider the homogeneous case.

THEOREM 4.22   *The solutions of the system of linear equations*

$$\sum_{j=1}^{n} \alpha_{ij} x_j = 0 \qquad (i = 1, \ldots, m)$$

form a vector space of dimension n-rank A, where $A = (\alpha_{ij}) \in M_{m,n}(K)$.
A non-trivial solution exists if and only if $n > rank\ A$. If $n = rank\ A$,
the trivial solution is the only solution.

PROOF   In the above notation, finding a solution of the above system
is equivalent to finding a $v = (x_1, x_2, \ldots, x_n) \in V_n(K)$ such that $Tv = 0$,
i.e. $v \in V_n(K)$ is a solution if and only if $v \in \ker T$. Thus the solution
space is $\ker T$ which is a subspace of $V_n(K)$. But by Theorem 4.9

$$rank\ T + nullity\ T = (V_n : K) = n$$

and thus

$$nullity\ T = n - rank\ T$$
$$= n - rank\ A$$

Hence, the system has a non-trivial solution if and only if
$n - rank\ A > 0$.
If $n = rank\ A$ then $\ker T = \{0\}$ and the trivial solution is unique.   ▋
   We now consider the non-homogeneous case, i.e.

$$\sum_{j=1}^{n} \alpha_{ij} x_j = \beta_i \qquad (i = 1, \ldots, m)$$

A solution of this system exists if and only if $(\beta_1, \ldots, \beta_m) \in im\ T$. But
the columns of A generate im T, thus a solution exists if and only if
$(\beta_1, \ldots, \beta_m)$ is a linear combination of the columns of A.
   Let $(A|b)$ be the augmented matrix of A, i.e. the matrix obtained by
adjoining $b = (\beta_1, \ldots, \beta_m)^t$ to the matrix A. Then we can prove

THEOREM 4.23   *A solution of the system of linear equations*

$$\sum_{j=1}^{n} \alpha_{ij} x_j = \beta_i \qquad (i = 1, \ldots, m)$$

exists if and only if the rank $(A|b) = rank\ A$.

PROOF   A solution of the system exists if and only if $(\beta_1, \beta_2, \ldots, \beta_m)$
is a linear combination of the columns of A.
   Thus, if a solution exists, rank $(A|b) = rank\ A$. Conversely, if
rank $(A|b) = rank\ A$, then either $b = 0$ or b is a linear combination
of the columns of A and so a solution exists.   ▋

The results of Theorem 4.22 and 4.23 should be compared with the criteria given for the solution of linear equations in Chapter 1.

EXAMPLES

1. Do the following systems of linear equations have non-trivial solutions?

(i)
$$x_1 + 2x_2 - x_3 + x_4 = 0$$
$$2x_1 - x_2 + x_3 + x_4 = 0$$
$$x_1 - x_2 + x_3 + 2x_4 = 0$$

(ii)
$$x_1 + 2x_2 - x_3 = 0$$
$$2x_1 - x_2 + x_3 = 0$$
$$3x_1 + x_2 + x_3 = 0$$

(i)
$$A = \begin{pmatrix} 1 & 2 & -1 & 1 \\ 2 & -1 & 1 & 1 \\ 1 & -1 & 1 & 2 \end{pmatrix}$$

Clearly, the rank $A \leqslant 3 < 4$ and by Theorem 4.22 the system has a non-trivial solution.

(ii)
$$A = \begin{pmatrix} 1 & 2 & -1 \\ 2 & -1 & 1 \\ 3 & 1 & 1 \end{pmatrix}$$

which is easily shown to be row equivalent to $\begin{pmatrix} 1 & 0 & 0 \\ 0 & 1 & 0 \\ 0 & 0 & 1 \end{pmatrix}$, i.e. the

rank $A = 3$ and again by Theorem 4.22 the trivial solution is unique.

2. Solve, if a solution exists, the system of linear equations

$$x_1 + x_2 - 2x_3 + x_4 + 3x_5 = 1$$
$$2x_1 - x_2 + 2x_3 + 2x_4 + 6x_5 = 2$$
$$3x_1 + 2x_2 - 4x_3 - 3x_4 - 9x_5 = 3$$

The augmented matrix is

$$(A|b) = \begin{pmatrix} 1 & 1 & -2 & 1 & 3 & | & 1 \\ 2 & -1 & 2 & 2 & 6 & | & 2 \\ 3 & 2 & -4 & -3 & -9 & | & 3 \end{pmatrix}$$

which is row equivalent to

$$\begin{pmatrix} 1 & 0 & 0 & 0 & 0 & | & 1 \\ 0 & 1 & -2 & 0 & 0 & | & 0 \\ 0 & 0 & 0 & 1 & 3 & | & 0 \end{pmatrix}$$

124

Since we clearly have in this case that rank $(A|b) = $ rank A a solution exists (by Theorem 4.23). A solution is now found by the methods given in Chapter I, i.e. from the above, we see that the system reduces to

$$x_1 = 1$$
$$x_2 - 2x_3 = 0$$
$$x_4 + 3x_5 = 0$$

and a general solution is of the form

$$(1, 2\lambda, \lambda, -3\mu, \mu) = (1, 0, 0, 0, 0) + \lambda(0, 2, 1, 0, 0) + \mu(0, 0, 0, -3, 1)$$

3. For the system of linear equations

$$x_1 + x_2 + 2x_3 + x_4 = 5$$
$$2x_1 + 3x_2 - x_3 - 2x_4 = 2$$
$$4x_1 + 5x_2 + 3x_3 = 7$$

the augmented matrix is

$$(A|b) = \begin{pmatrix} 1 & 1 & 2 & 1 & | & 5 \\ 2 & 3 & -1 & -2 & | & 2 \\ 4 & 5 & 3 & 0 & | & 7 \end{pmatrix}$$

which is row equivalent to

$$\begin{pmatrix} 1 & 0 & 7 & 5 & | & 13 \\ 0 & 1 & 5 & -4 & | & 8 \\ 0 & 0 & 0 & 0 & | & -5 \end{pmatrix}$$

which implies that rank $(A|b) > $ rank A and so the system does not have a solution.

Exercises 4.6

1. Show that the rank of the matrix

$$\begin{pmatrix} 1 - \sqrt{6} & \sqrt{3} & \sqrt{2} \\ 2 & \sqrt{6} & -\sqrt{2} \\ 1 & -\sqrt{3} & \sqrt{2} - \sqrt{6} \end{pmatrix}$$

is 2.

2. Find the rank of the following matrices for all values of a

(i) $\begin{pmatrix} 1 & 3 & 5 & 6+a \\ 2 & 3 & 4-a & 2 \\ 1 & 1-a & -2 & -5 \\ 1 & 6 & 12 & 19 \end{pmatrix}$ (ii) $\begin{pmatrix} 1 & 1-2a & 1 & 1-2a \\ 1+2a & -1-a & -1+2a & 1-a \\ 2-a & 3a & 2a & 2-a \\ 5a & 1-a & 1+3a & 0 \end{pmatrix}$

3. Show that the rank of the matrix

$$\begin{pmatrix} a & b & b & a \\ b & a & -a & -b \\ a+b & a+b & 2a & -2a \\ -2a & 2a & a+b & a+b \end{pmatrix}$$

is 4 unless $a+b=0$ or $b=3a$. Find the rank in each of these cases.

4. Find all values of t for which the matrix

$$\begin{pmatrix} (1+t)t & t-1 & -t \\ 0 & 2 & -1 \\ -2t & 4-2t & t-2 \end{pmatrix}$$

is of rank less than 3 and determine the corresponding rank. In each case express one of the columns as a linear combination of the others.

5. Find the rank of the linear transformation defined in Exercises 4.2, No. 7 for all values of α, β and γ.

6. Let $S \in M_n(K)$ be fixed and T be the linear transformation on $M_n(K)$ defined by $T(A)=AS$. If S is an invertible matrix, show that rank $T=n^2$. In general, prove that rank $T=n$ rank S.

CHAPTER 5

Inner Product Spaces

5.1 Introduction and Three-Dimensional Geometry

In Chapter III, vector spaces over any field were defined as generalizations of 2- and 3-dimensional real spaces and many of the basic concepts in these cases were extended to vector spaces in general, e.g. K-basis, linear transformation etc. But there are in addition, further concepts which prove to be useful in these cases, namely the length of a line, angle between two lines, perpendicular lines, etc. This chapter will consider the generalization of these ideas to vector spaces over the real and complex fields. Before considering this generalization, we shall review these concepts in 3-space $V_3(\mathbf{R})$.

Let $v = (\alpha, \beta, \gamma) \in V_3(\mathbf{R})$, then refering to Figure 1, the **length** of v, which is denoted by $\|v\|$ is

$$
\begin{aligned}
\|v\| &= \sqrt{OR^2 + PR^2} \\
&= \sqrt{OQ^2 + QR^2 + PR^2} \\
&= \sqrt{\alpha^2 + \beta^2 + \gamma^2}
\end{aligned}
$$

where $\sqrt{}$ means the positive square root.
Note that if $\lambda \in \mathbf{R}$ then

$$\|\lambda v\| = |\lambda| \, \|v\|$$

since $\|\lambda v\| = \sqrt{\lambda^2 (\alpha^2 + \beta^2 + \gamma^2)}$

$$= |\lambda| \, \|v\|$$

$v \in V_3(\mathbf{R})$ is called a **unit** vector if $\|v\| = 1$. For example,

$(1, 0, 0), (0, 1, 0), (0, 0, 1), \left(\dfrac{1}{\sqrt{2}}, \dfrac{1}{\sqrt{2}}, 0\right), \left(\dfrac{1}{\sqrt{3}}, \dfrac{1}{\sqrt{3}}, \dfrac{1}{\sqrt{3}}\right)$ are unit

vectors in $V_3(\mathbf{R})$. If $v \in V_3(\mathbf{R})$ then $\dfrac{1}{\|v\|} v$ is a unit vector.

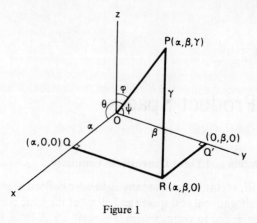

Figure 1

Let θ, ψ, ϕ be the angles which OP makes with the x-, y- and z-axes respectively, then $\cos\theta = \dfrac{OQ}{OP} = \dfrac{\alpha}{\|v\|}$, $\cos\psi = \dfrac{\beta}{\|v\|}$, $\cos\phi = \dfrac{\gamma}{\|v\|}$. $\cos\theta$, $\cos\psi$ and $\cos\phi$ are called the **direction cosines** of v.

Now, let v and w be two vectors in $V_3(\mathbf{R})$, represented by the points P and Q respectively, see Fig. 2.

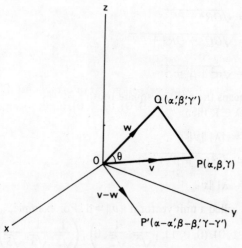

Figure 2

The **distance** between P and Q, denoted by $\mathrm{d}(P, Q)$, or the length of PQ is equal to the length of OP', where P' is the point $(\alpha - \alpha', \beta - \beta', \gamma - \gamma')$. Thus

128

$$d(P, Q) = \|(\alpha - \alpha', \beta - \beta', \gamma - \gamma')\|$$
$$= \sqrt{(\alpha - \alpha')^2 + (\beta - \beta')^2 + (\gamma - \gamma')^2}$$
$$= \|v - w\|$$

The **angle** between v and w is defined to be the angle $POQ = \theta$ such that $0 \leqslant \theta \leqslant \pi$. Now, by the cosine rule

$$PQ^2 = OP^2 + OQ^2 - 2.OP.OQ \cos \theta$$

that is,

$$\cos \theta = \frac{\|v - w\|^2 - \|v\|^2 - \|w\|^2}{-2 \|v\| \|w\|}$$

$$= \frac{(\alpha - \alpha')^2 + (\beta - \beta')^2 + (\gamma - \gamma')^2 - (\alpha^2 + \beta^2 + \gamma^2 + \alpha'^2 + \beta'^2 + \gamma'^2)}{-2 \|v\| \|w\|}$$

$$= \frac{\alpha\alpha' + \beta\beta' + \gamma\gamma'}{\|v\| \|w\|}$$

If $v = (\alpha, \beta, \gamma)$, $w = (\alpha', \beta', \gamma')$, then the **inner product** (or **dot product**) of v and w, denoted by (v, w) (or $v.w$) is defined by

$$(v, w) = \alpha\alpha' + \beta\beta' + \gamma\gamma'$$

In that case, from the above, we have

$$(v, w) = \|v\| \|w\| \cos \theta$$

which is sometimes given as the definition of inner product. Note also that $\|v\| = \sqrt{(v, v)}$.

Two vectors v and w are **perpendicular** or **orthogonal** if $\cos \theta = 0$ or $(v, w) = 0$. For example $(1, 0, 0)$, $(0, 1, 0)$, $(0, 0, 1)$ are mutually perpendicular. Also $\left(\frac{1}{\sqrt{2}}, 0, \frac{1}{\sqrt{2}}\right)$ and $\left(\frac{1}{\sqrt{2}}, 0, -\frac{1}{\sqrt{2}}\right)$ are perpendicular to each other.

In physical applications the vectors $(1, 0, 0)$, $(0, 1, 0)$ and $(0, 0, 1)$ are denoted by $\mathbf{i}$, $\mathbf{j}$, and $\mathbf{k}$, respectively and as we saw in Chapter III, $\{\mathbf{i}, \mathbf{j}, \mathbf{k}\}$ is the standard $\mathbf{R}$-basis for $V_3(\mathbf{R})$. Thus, $V_3(\mathbf{R})$ has an $\mathbf{R}$-basis consisting of vectors which are mutually perpendicular or orthogonal. This is the foundation for the Cartesian coordinate system which is of fundamental importance in the development of geometry. In the next section, having generalized the concept of orthogonal vectors to arbitrary vector space, we shall show that every real and complex finite dimensional vector space has a basis consisting of orthogonal vectors.

The following theorem can be proved concerning the inner product

THEOREM 5.1   *If u, v, w $\in V_3(\mathbf{R})$ and $\lambda \in \mathbf{R}$ then*

(i) $(u + v, w) = (u, w) + (v, w)$

(ii) $(\lambda v, w) = \lambda(v, w)$

(iii) $(v, w) = (w, v)$

(iv) $(v, v) > 0$ if $v \neq 0$.

PROOF   The proof is elementary, for example
if $v = (\alpha, \beta, \gamma)$, $w = (\alpha', \beta', \gamma')$ then

$$(\lambda v, w) = \lambda\alpha\alpha' + \lambda\beta\beta' + \lambda\gamma\gamma'$$

$$= \lambda(\alpha\alpha' + \beta\beta' + \gamma\gamma')$$

$$= \lambda(v, w)$$

proving (ii). $\alpha^2 + \beta^2 + \gamma^2 > 0$ with $\alpha, \beta, \gamma \in \mathbf{R}$ if and only if at least
one of α, β or γ is non-zero, that is $(\alpha, \beta, \gamma) \neq (0, 0, 0)$, which proves (iv). ∎

When vector spaces were first defined in Chapter 3, the basic properties
verified in the special case $V_n(K)$ were taken as motivation for the
definition of abstract vector spaces. For the new concepts which are to
be introduced now, it is the properties of the inner product enunciated
in Theorem 5.1 which are crucial to make things work. Thus, we again
reverse the process and take the statements in Theorem 5.1 as the basis
of definition of inner product spaces given in the next section.
Attention is restricted to the case where the field K is a real or complex
field (or their subfields). Statement (iv) will only be meaningful if (v, v)
takes values in an ordered field such as the real field. By modifying the
requirement (iii), this can also be assured when K is the complex field.

It would be possible to develop the theory of vector spaces over an
arbitrary field with a symmetric inner product, i.e. an inner product
satisfying (i), (ii) and (iii) but it will be seen that (iv) is absolutely
essential if the crucial concepts of the length of a vector, distance
between points, the angle between two vectors and perpendicular or
orthogonal vectors are to be generalized.

Exercises 5.1

1. Which of the following pairs of vectors are perpendicular?

(i) $(2, -1, 1)$ and $(1, 2, 1)$

(ii) $(2, 1, -3)$ and $(1, 1, 1)$

(iii) $(7, 5, 3)$ and $(1, -2, 1)$.

2. Find a vector perpendicular to $(2, -1, 2)$ and $(1, -1, 2)$.

130

3. Find the lengths of the following vectors
   (i) $(1, 2, 1)$,   (ii) $(3, -2, 5)$,   (iii) $(1, 0, -1)$.

4. Find the angle between the following pairs of vectors
   (i) $(3, -2, 1)$ and $(1, -1, 1)$
   (ii) $(2, 1, -1)$ and $(1, 0, 2)$.

5.2 Euclidean and Unitary Spaces

Throughout this section K will stand for either the field of real numbers $\mathbf{R}$ or the field of complex numbers $\mathbf{C}$ and V a K-space.

DEFINITION 5.2   *An **inner product** on V is a function which assigns to each ordered pair of vectors u, $v \in V$ a scalar $(u, v) \in K$ with the following properties*
   (i) $(u + w, v) = (u, v) + (w, v)$ *for all* u, v, $w \in V$
   (ii) $(\alpha u, v) = \alpha(u, v)$ *for all* u, $v \in V$, $\alpha \in K$
   (iii) $(v, u) = \overline{(u, v)}$, *for all* u, $v \in V$
   (iv) $(v, v) > 0$ *if* $v \neq 0$,
*where the bar denotes taking the complex conjugate. A vector space V with inner product is called an **inner product space**. In particular, a real inner product space is called a **Euclidean space** and a complex inner product space is called a **unitary space**.*

Before looking at examples, we note that (i), (ii) and (iii) imply that

$$(\alpha u + \beta w, v) = \alpha(u, v) + \beta(w, v)$$

and

$$(u, \alpha v + \beta w) = \bar{\alpha}(u, v) + \bar{\beta}(u, w)$$

for all $\alpha, \beta \in K, u, v, w \in V$, since

$$(\alpha u + \beta w, v) = (\alpha u, v) + (\beta w, v) \qquad \text{by (i)}$$
$$= \alpha(u, v) + \beta(w, v) \qquad \text{by (ii)}$$

and

$$(u, \alpha v + \beta w) = \overline{(\alpha v + \beta w, u)} \qquad \text{by (iii)}$$
$$= \overline{\alpha(v, u) + \beta(w, u)} \qquad \text{by the above}$$
$$= \bar{\alpha}(u, v) + \bar{\beta}(u, w) \qquad \text{by (iii)}.$$

Note also that (iii) ensures that $(v, v) \in \mathbf{R}$ and so (iv) is a meaningful statement.

EXAMPLES

1. If $u = (\alpha_1, \alpha_2, \ldots, \alpha_n)$, $v = (\beta_1, \beta_2, \ldots, \beta_n) \in V_n(\mathbf{R})$, define

$$(u, v) = \alpha_1 \beta_1 + \alpha_2 \beta_2 + \ldots + \alpha_n \beta_n$$

then it is easily shown that this is an inner product on $V_n(\mathbf{R})$ and thus $V_n(\mathbf{R})$ is a Euclidean space. This is a natural generalization of the inner product on $V_3(\mathbf{R})$ considered in Section 1, and will be called the **standard** inner product on $V_n(\mathbf{R})$.

2. If $u = (\alpha_1, \alpha_2, \ldots, \alpha_n)$, $v = (\beta_1, \beta_2, \ldots, \beta_n) \in V_n(\mathbf{C})$, define

$$(u, v) = \alpha_1 \bar{\beta}_1 + \alpha_2 \bar{\beta}_2 + \ldots + \alpha_n \bar{\beta}_n$$

then again it is easily shown that $V_n(\mathbf{C})$ is an inner product space and thus a unitary space. This will be called the **standard** inner product on $V_n(\mathbf{C})$.

3. If $f, g \in C[a, b]$, define

$$(f, g) = \int_a^b f(t)\, g(t)\, \mathrm{d}t$$

then (i)-(iv) in Definition 5.2 are familiar properties of integration. This will be called the **standard** inner product on $C[a, b]$.

4. If $u = (\alpha_1, \alpha_2, \ldots, \alpha_n)$, $v = (\beta_1, \beta_2, \ldots, \beta_n) \in V_n(\mathbf{R})$, define

$$(u, v) = \alpha_1 \beta_1 + 2\alpha_2 \beta_2 + \ldots + n\alpha_n \beta_n$$

then it is again easily verified that this also defines an inner product on $V_n(\mathbf{R})$.

From now on in this chapter V will denote an inner product space.

DEFINITION 5.3 *The **length** or **norm** of $v \in V$ denoted by $\|v\|$ is defined by*

$$\|v\| = \sqrt{(v, v)}$$

*A vector $v \in V$ with $\|v\| = 1$ is called a **unit vector**.*

Note that by Definition 5.2 (iv) if $v \neq 0$ $(v, v) > 0$ and so $\sqrt{(v, v)}$ is a positive real number. Also if $v \in V$, $v \neq 0$ then $\dfrac{1}{\|v\|} v$ is a unit vector; we say that the vector v has been **normalized**.

In the case of $V_n(\mathbf{R})$, this is clearly a generalization of length in $V_3(\mathbf{R})$ considered in §5.1. In $V_n(\mathbf{C})$, if $v = (\alpha_1, \alpha_2, \ldots, \alpha_n) \in V_n(\mathbf{C})$, then

132

$$(v, v) = \alpha_1 \bar{\alpha}_1 + \alpha_2 \bar{\alpha}_2 + \ldots + \alpha_n \bar{\alpha}_n$$
$$= |\alpha_1|^2 + |\alpha_2|^2 + \ldots + |\alpha_n|^2$$

The next theorem shows that the length of a vector has some of the familiar properties of the length of a line in the plane or in 3-space.

THEOREM 5.4 *If $u, v \in V$ and $\alpha \in K$, then*
(i) $\|\alpha v\| = |\alpha| \, \|v\|$,
(ii) $\|v\| > 0$ *if* $v \neq 0$,
(iii) $|(u, v)| \leqslant \|u\| \, \|v\|$
(iv) $\|u + v\| \leqslant \|u\| + \|v\|$.

PROOF (i) $\|\alpha v\| = (\alpha v, \alpha v)^{\frac{1}{2}}$
$$= (\alpha \bar{\alpha}(v, v))^{\frac{1}{2}}$$
$$= |\alpha| \, \|v\|.$$

(ii) follows immediately from (iv) of Definition 5.2.

(iii) if $u = 0$, then both sides are 0 and the result holds.

If $u \neq 0$, put $w = v - \dfrac{(v, u)}{\|u\|^2} u$, then

$$(w, u) = (v, u) - \frac{(v, u)}{\|u\|^2}(u, u)$$
$$= 0$$

and

$$0 \leqslant \|w\|^2 = (w, w)$$
$$= \left(w, v - \frac{(v, u)}{\|u\|^2}u\right)$$
$$= (w, v)$$
$$= (v, v) - \frac{(v, u)}{\|u\|^2}(u, v)$$
$$= \|v\|^2 - \frac{|(u, v)|^2}{\|u\|^2}$$

that is

$$|(u, v)|^2 \leqslant \|u\|^2 \, \|v\|^2$$

or

$$|(u, v)| \leqslant \|u\| \, \|v\|$$

133

(iv)   Now, by means of (iii) above, we have

$$\|u + v\|^2 = (u+v, u+v) = \|u\|^2 + (u, v) + (v, u) + \|v\|^2$$
$$\leqslant \|u\|^2 + |(u,v) + (v,u)| + \|v\|^2$$
$$\leqslant \|u\|^2 + |(u,v)| + |(v,u)| + \|v\|^2$$
$$\leqslant \|u\|^2 + 2\|u\|\,\|v\| + \|v\|^2$$
$$= (\|u\| + \|v\|)^2$$

and hence

$$\|u + v\| \leqslant \|u\| + \|v\|$$

(iii) is called the **Cauchy-Schwarz inequality** and is a generalization of familiar inequalities in other settings as we see below and (iv) is called the **triangle inequality**, for in $V_3(\mathbf{R})$ it reduces to the well-known statement that the length of a side of a triangle is less than the sum of the other two sides as is seen in Fig. 3.

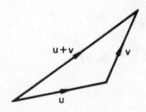

Figure 3

Alternatively, we define

$$d(u, v) = \|v - u\|$$

for all $u, v \in V$, and call $d(u, v)$ the **distance** between u and v. Then we have an alternative version of the triangle inequality.

COROLLARY   *If $u, v, w \in V$ then*

$$d(u, w) \leqslant d(u, v) + d(v, w)$$

PROOF   Replace u and v in (iv) above by $v - u$ and $w - v$ respectively, then $u + v$ is replaced by $w - u$ and the statement (iv) of Theorem 5.4 now becomes

$$d(u, w) \leqslant d(u, v) + d(v, w)$$

134

In $V_n(\mathbf{R})$ and $V_n(\mathbf{C})$ with standard inner products (iii) becomes

$$(\alpha_1\beta_1 + \alpha_2\beta_2 + \ldots + \alpha_n\beta_n)^2 \leqslant (\alpha_1{}^2 + \alpha_2{}^2 + \ldots + \alpha_n{}^2)$$
$$(\beta_1{}^2 + \beta_2{}^2 + \ldots + \beta_n{}^2)$$

and

$$|(\alpha_1\beta_1 + \alpha_2\beta_2 + \ldots + \alpha_n\beta_n)|^2$$
$$\leqslant (|\alpha_1|^2 + |\alpha_2|^2 + \ldots + |\alpha_n|^2)(|\beta_1|^2 + |\beta_2|^2 + \ldots + |\beta_n|^2)$$

respectively. These two statements (the first is obviously a special case of the second) are called the Cauchy inequalities.

In $C[a,b]$ with the standard inner product, (iii) becomes

$$\left(\int_a^b f(t)g(t)\,\mathrm{d}t\right)^2 \leqslant \left(\int_a^b f(t)^2\,\mathrm{d}t\right)\left(\int_a^b g(t)^2\,\mathrm{d}t\right)$$

which again is an important property of integration.

The concept of angle can also be introduced for real Euclidean spaces V. If $u, v \in V$, then by Theorem 5.4 (iii), we have

$$\frac{|(u, v)|}{\|u\|\,\|v\|} \leqslant 1$$

or, what is the same,

$$-1 \leqslant \frac{(u, v)}{\|u\|\,\|v\|} \leqslant 1$$

and $(u, v)/\|u\|\,\|v\|$ is a real number. The **angle** θ between u and v is then defined to be that number $0 \leqslant \theta \leqslant \pi$ such that

$$\cos\theta = \frac{(u, v)}{\|u\|\,\|v\|}$$

what has been said above ensuring that θ is well defined and also that this definition is meaningful, i.e. $\cos\theta$ is a real number and $-1 \leqslant \cos\theta \leqslant 1$ for all angles θ. It is clear that this definition of angle cannot be extended to unitary spaces. If $\theta = \pi/2$, then $(u, v) = 0$ and we say that u and v are **orthogonal** or **perpendicular** to each other. This will be subject of the next section; since the statement "$(u, v) = 0$" is meaningful for unitary spaces also, the concept of orthogonal vectors can be introduced for arbitrary inner product spaces.

EXAMPLE

In $V_6(\mathbf{R})$ with standard inner product, if $u = (3, -2, -3, 1, 1, -1)$, $v = (-1, 0, 0, 1, 1, 1)$, find the lengths of u and v and the angle between u and v. Verify (iv) of Theorem 5.4 in this case.

By Definition 5.3, we have

$$\|u\|^2 = (u, u) = 9 + 4 + 9 + 1 + 1 + 1 = 25$$

$$\|v\|^2 = (v, v) = 1 + 1 + 1 + 1 \qquad = 4$$

thus $\|u\| = 5$, $\|v\| = 2$.
The angle θ between u and v is given by

$$\cos \theta = \frac{-3 + 1 + 1 - 1}{5.2} = \frac{-2}{5.2} = \frac{-1}{5}$$

thus $\theta = \cos^{-1}(-1/5) \, (\cong 101° \, 34') \; u + v = (2, -2, -3, 2, 2, 0)$ and thus

$$\|u + v\| = \sqrt{4 + 4 + 9 + 4 + 4} = \sqrt{21} < 2 + 5 = \|u\| + \|v\|$$

Exercises 5.2

1. Which of the following are inner products on $V_2(\mathbf{R})$, if $u = (\alpha_1, \alpha_2)$, $v = (\beta_1, \beta_2)$?
   - (i) $(u, v) = \alpha_1\beta_1 - \alpha_2\beta_1 - \alpha_1\beta_2 + 2\alpha_2\beta_2$
   - (ii) $(u, v) = \alpha_1\beta_1 + \alpha_2\beta_1 + \alpha_1\beta_2 - \alpha_2\beta_2$
   - (iii) $(u, v) = \alpha_1\beta_1 - \alpha_2\beta_1 + \alpha_1\beta_2 + 2\alpha_2\beta_2$
   - (iv) $(u, v) = \alpha_1^2\beta_1^2 + \alpha_2^2\beta_2^2$.

2. Which of the following are inner products on $V_3(\mathbf{R})$, if $u = (\alpha_1, \alpha_2, \alpha_3)$, $v = (\beta_1, \beta_2, \beta_3)$?
   - (i) $(u, v) = \alpha_1\beta_1 + 2\alpha_2\beta_2 + 3\alpha_3\beta_3 + \alpha_1\beta_2 + \alpha_2\beta_1 + \alpha_1\beta_3$
     $+ \alpha_3\beta_1 + 2\alpha_2\beta_3 + 2\alpha_3\beta_2$
   - (ii) $(u, v) = \alpha_1\beta_1 + 2\alpha_2\beta_2 + 3\alpha_3\beta_3 + 2\alpha_1\beta_2 + 2\alpha_1\beta_3 + 4\alpha_2\beta_3$
   - (iii) $(u, v) = \alpha_1\beta_1 + \alpha_3\beta_3 - \alpha_1\beta_2 - \alpha_2\beta_1 - \alpha_1\beta_3 - \alpha_3\beta_1$
     $+ \alpha_2\beta_3 + \alpha_3\beta_2$.

3. Which of the following are inner products on $C[1, -1]$, the vector space of real valued continuous functions defined on $[-1, 1]$, if $f, g \in C[1, -1]$?
   - (i) $(f, g) = \int_{-1}^{1} f(x) g(x) \, dx$

   - (ii) $(f, g) = \int_{-1}^{1} (1 - x^2) f(x) g(x) \, dx$

   - (iii) $(f, g) = \int_{-1}^{1} x^2 f(x) g(x) \, dx$.

4. Which of the following are inner products on $M_n(\mathbf{R})$, if $A, B \in M_n(\mathbf{R})$,
   (i) $(A, B) = \text{trace } (AB)$
   (ii) $(A, B) = \det (AB)$.

5. Compute $\|u\|, \|v\|, \|u + v\|, (u, v)$ and the angle between u and v and verify that the Cauchy-Schwarz and Triangle Inequalities hold if
   (i) $u = (1, 0, 2, -2)$ and $v = (2, 1, -2, 0)$ are elements of $V_4(\mathbf{R})$ with standard inner product;
   (ii) $u = (1, 0, 2)$ and $v = (2, 1, 2)$ are elements of $V_3(\mathbf{R})$ for the inner products defined in Exercise 2 above.
   (iii) $u = x$ and $v = \cos \pi x$ are elements of $C[0, 1]$ with the standard inner product.

6. If V is a Euclidean space and $u, v \in V$ prove that
   (i) $|\|u\| - \|v\|| \leqslant \|u - v\|$
   (ii) $4(u, v) = \|u + v\|^2 - \|u - v\|^2$
   (iii) $\|u - v\|^2 + \|u + v\|^2 = 2(\|u\|^2 + \|v\|^2)$.
What does (iii) say about the diagonals of a parallelogram?

7. If $u, v \in V$, show that the distance function $d(u, v)$ satisfies
   (i) $d(u, v) \geqslant 0$
   (ii) $d(u, v) = d(v, u)$
   (iii) $d(u, v) = 0$ if and only if $u = v$.

5.3 Orthogonal Vectors

Let V be an inner product space.

DEFINITION 5.5   *If $u, v \in V$ and $(u, v) = 0$ then u and v are said to be **orthogonal** (or **perpendicular**) to each other. A subset S of V is called an **orthogonal set** if the elements of S are mutually orthogonal. An orthogonal set is called an **orthonormal set** if each vector has unit length i.e. $\|v\| = 1$.*

EXAMPLES

1. In $V_6(\mathbf{R})$ with standard inner product, find the vectors orthogonal to $v = (3, -2, -3, 1, 1, -1)$.

   If $u = (\alpha_1, \alpha_2, \alpha_3, \alpha_4, \alpha_5, \alpha_6) \in V_6(\mathbf{R})$ is orthogonal to v then

   $$3\alpha_1 - 2\alpha_2 - 3\alpha_3 + \alpha_4 + \alpha_5 - \alpha_6 = 0$$

   that is, the set of all vectors u orthogonal to v is given by all the solutions to this linear equation.

Clearly $\{(1, 0, 0, 0, 0, 3), (0, 1, 0, 0, 0, -2), (0, 0, 1, 0, 0, -3),$
$(0, 0, 0, 1, 0, 1), (0, 0, 0, 0, 1, 1)\}$
is an **R**-basis for the solution space to this linear equation and all
R-linear combinations of these vectors are orthogonal to v.

2. The standard bases for $V_n(\mathbf{R})$ and $V_n(\mathbf{C})$ are orthonormal relative
to the standard inner product.

3. In $C[1, -1]$, prove that $\{P_0(x) = 1, P_1(x) = x, P_2(x) = \frac{1}{2}(3x^2 - 1)\}$
is an orthogonal set of vectors. Find the length of each element.

$$(P_0(x), P_1(x)) = \int_{-1}^{1} x \, dx = \left[\frac{x^2}{2}\right]_{-1}^{1} = \frac{1}{2} - \frac{1}{2} = 0$$

$$(P_0(x), P_2(x)) = \int_{-1}^{1} \frac{1}{2}(3x^2 - 1) \, dx = \left[\frac{x^3}{2} - \frac{x}{2}\right]_{-1}^{1}$$

$$= \frac{1}{2} - \frac{1}{2} - \frac{1}{2} + \frac{1}{2} = 0,$$

$$(P_1(x), P_2(x)) = \int_{-1}^{1} \frac{1}{2} x (3x^2 - 1) \, dx = \left[\frac{3x^4}{8} - \frac{x^2}{4}\right]_{-1}^{1}$$

$$= \frac{3}{8} - \frac{1}{4} - \frac{3}{8} + \frac{1}{4} = 0.$$

$$(P_0(x), P_0(x)) = \int_{-1}^{1} dx = [x]_{-1}^{1} = 1 + 1 = 2,$$

$$(P_1(x), P_1(x)) = \int_{-1}^{1} x^2 \, dx = \left[\frac{x^3}{3}\right]_{-1}^{1} = \frac{1}{3} + \frac{1}{3} = \frac{2}{3},$$

$$(P_2(x), P_2(x)) = \int_{-1}^{1} \frac{1}{4}(3x^2 - 1)^2 \, dx = \left[\frac{9x^5}{20} - \frac{6x^3}{12} + \frac{1}{4}x\right]_{-1}^{1} = \frac{2}{5},$$

thus $P_0(x)$, $P_1(x)$ and $P_2(x)$ have lengths $\sqrt{2}$, $\sqrt{\frac{2}{3}}$, and $\sqrt{\frac{2}{5}}$
respectively.

LEMMA 5.6   *An orthogonal set of non-zero vectors in an inner
product space V is linearly independent.*

PROOF   Let $\{v_1, v_2, \ldots, v_n\}$ be an orthogonal set of non-zero
vectors in V. Consider

$$\alpha_1 v_1 + \alpha_2 v_2 + \ldots + \alpha_n v_n = 0$$

where $\alpha_i \in K$ $(i = 1, 2, \ldots, n)$. Then for $1 \leqslant k \leqslant n$,

138

$$0 = \left(\sum_{i=1}^{n} \alpha_i v_i, v_k \right) = \sum_{i=1}^{n} \alpha_i (v_i, v_k)$$

$$= \alpha_k (v_k, v_k)$$

and since $v_k \neq 0$, $(v_k, v_k) \neq 0$ and so $\alpha_k = 0$.

Thus $\{v_1, v_2, \ldots, v_n\}$ is linearly independent over K. ∎

We now prove our main result which shows that every finite dimensional inner product space has a basis consisting of orthonormal vectors. Furthermore, the proof is constructive in that it gives a procedure for determining an orthonormal basis from any given basis for V.

THEOREM 5.7 (*Gram-Schmidt Orthogonalization Procedure*).
Every finite dimensional inner product space has a basis consisting of orthonormal vectors.

PROOF Let $\{v_1, v_2, \ldots, v_n\}$ be a K-basis for the inner product space V. Define the subset $\{u_1, u_2, \ldots, u_n\}$ of V inductively as follows:

$$u_1 = v_1$$

$$u_2 = v_2 - \frac{(v_2, u_1)}{\|u_1\|^2} u_1$$

$$u_3 = v_3 - \frac{(v_3, u_2)}{\|u_2\|^2} u_2 - \frac{(v_3, u_1)}{\|u_1\|^2} u_1$$

$$\vdots$$

$$u_n = v_n - \frac{(v_n, u_{n-1})}{\|u_{n-1}\|^2} u_{n-1} - \ldots - \frac{(v_n, u_1)}{\|u_1\|^2} u_1$$

that is, the coefficient of u_j in u_i $j < i$ is the inner product of v_i with u_j divided by the length of u_j.

Then $u_1, u_2, \ldots, u_n$ are non-zero, otherwise we contradict the linear independence of $\{v_1, \ldots, v_n\}$. We show that $\{u_1, u_2, \ldots, u_n\}$ is orthogonal by induction on n. If $n = 2$, then

$$(u_2, u_1) = (v_2, u_1) - \frac{(v_2, u_1)}{\|u_1\|^2}(u_1, u_1) = 0$$

and so $\{u_1, u_2\}$ is an orthogonal set.

If $n > 2$, assume that $\{u_1, u_2, \ldots, u_{n-1}\}$ is an orthogonal set. To show that $\{u_1, u_2, \ldots, u_n\}$ is an orthogonal set, we must show that $(u_n, u_i) = 0$ $(i = 1, 2, \ldots, n-1)$. Now, for $i = 1, 2, \ldots, n-1$, we have by the induction assumption that

$$(u_n, u_i) = (v_n, u_i) - \frac{(v_n, u_{n-1})}{\|u_{n-1}\|^2}(u_{n-1}, u_i) - \ldots - \frac{(v_n, u_1)}{\|u_1\|^2}(u_1, u_i)$$

$$= (v_n, u_i) - \frac{(v_n, u_i)(u_i, u_i)}{\|u_i\|^2}$$

$$= 0$$

as required.

Furthermore, by the previous lemma, $\{u_1, u_2, \ldots, u_n\}$ is linearly independent over K and so is a K-basis for V. Now, put

$w_i = \dfrac{1}{\|u_i\|} u_i \ (i = 1, 2, \ldots, n)$ and each w_i is a unit vector and

$\{w_1, w_2, \ldots, w_n\}$ is an orthonormal basis for V. ∎

In §5.2, Example 2 we defined the standard inner product on $V_n(\mathbf{C})$; we now show that every inner product on a finite dimensional inner product space essentially takes this form.

If V is a finite dimensional inner product space then by the above theorem it has a K-basis $\{v_1, v_2, \ldots, v_n\}$ consisting of orthonormal vectors. If $u, v \in V$ then $u = \sum\limits_{i=1}^{n} \alpha_i v_i, v = \sum\limits_{i=1}^{n} \beta_i v_i \ (\alpha_i, \beta_i \in K)$ and

$$(u, v) = \left(\sum_{i=1}^{n} \alpha_i v_i, \sum_{i=1}^{n} \beta_i v_i \right)$$

$$= \sum_{i,j=1}^{n} \alpha_i \bar{\beta}_j (v_i, v_j)$$

$$= \sum_{i=1}^{n} \alpha_i \bar{\beta}_i$$

since $(v_i, v_j) = \delta_{ij} (i, j = 1, 2, \ldots, n)$.

EXAMPLES

1. Apply the Gram-Schmidt orthogonalization procedure to the vectors $v_1 = (1, 0, 1)$, $v_2 = (1, 0, -1)$, $v_3 = (0, 3, 4)$ to obtain an orthonormal basis for $V_3(\mathbf{R})$.

Put

$$u_1 = v_1 = (1, 0, 1)$$
$$u_2 = v_2 = (1, 0, -1)$$

$$u_3 = v_3 - \frac{(v_3, u_2)}{(u_2, u_2)} u_2 - \frac{(v_3, u_1)}{(u_1, u_1)} u_1$$

$$= (0, 3, 4) - \frac{(-4)}{2}(1, 0, -1) - \frac{4}{2}(1, 0, 1)$$

$$= (0, 3, 0)$$

then $\{u_1, u_2, u_3\}$ is an orthogonal set. The lengths of these vectors are $\sqrt{2}, \sqrt{2}, 3$ respectively so the required **R**-basis for $V_3(\mathbf{R})$ is

$$\left\{ \frac{1}{\sqrt{2}}(1, 0, 1), \frac{1}{\sqrt{2}}(1, 0, -1), (0, 1, 0) \right\}.$$

2. Extend the orthonormal set $\{\frac{1}{3}(2, 0, -1, 2), \frac{1}{3}(2, 1, 0, -2)\}$ to give an orthonormal basis for $V_4(\mathbf{R})$. It is clear that the set $\{v_1 = \frac{1}{3}(2, 0, -1, 2), v_2 = \frac{1}{3}(2, 1, 0, -2), v_3 = (1, 0, 0, 0),$ $v_4 = (0, 0, 0, 1)\}$ is an **R**-basis for $V_4(\mathbf{R})$. Put

$$u_1 = v_1$$

$$u_2 = v_2$$

$$u_3 = v_3 - \frac{(v_3, u_2)}{(u_2, u_2)} u_2 - \frac{(v_3, u_1)}{(u_1, u_1)} u_1$$

$$= (1, 0, 0, 0) - \frac{2}{3} \cdot \frac{1}{3}(2, 1, 0, -2) - \frac{2}{3} \cdot \frac{1}{3}(2, 0, -1, 2)$$

$$= \frac{1}{9}(1, -2, 2, 0)$$

$$u_4 = v_4 - \frac{(v_4, u_3)}{(u_3, u_3)} u_3 - \frac{(v_4, u_2)}{(u_2, u_2)} u_2 - \frac{(v_4, u_1)}{(u_1, u_1)} u_1$$

$$= (0, 0, 0, 1) - 0 - (-\frac{2}{3}) \cdot \frac{1}{3}(2, 1, 0, -2) - \frac{2}{3} \cdot \frac{1}{3}(2, 0, -1, 2)$$

$$= (0, \frac{2}{9}, \frac{2}{9}, \frac{1}{9})$$

After normalizing these vectors we find that
$\{\frac{1}{3}(2, 0, -1, 2), \frac{1}{3}(2, 1, 0, -2), \frac{1}{3}(1, -2, 2, 0), \frac{1}{3}(0, 2, 2, 1)\}$
is the required orthonormal basis.

3. Let V be the subspace of real polynomials of degree $\leqslant 3$ in $C[1, -1]$. Find an orthonormal basis for V.

$\{f_0 = 1, f_1 = x, f_2 = x^2, f_3 = x^3\}$ is an **R**-basis for V.

By the Gram-Schmidt Orthogonalization Procedure we obtain an orthogonal set of vectors $\{ g_0, g_1, g_2, g_3\}$ as follows:

$$g_0 = f_0 = 1$$

$$g_1 = f_1 - \frac{(f_1, f_0)}{(f_0, f_0)} f_0 = x - \frac{\int_{-1}^{1} x\,dx}{\int_{-1}^{1} dx} \cdot 1 = x$$

$$g_2 = f_2 - \frac{(f_2, g_1)}{(g_1, g_1)} g_1 - \frac{(f_2, g_0)}{(g_0, g_0)} g_0$$

$$= x^2 - \frac{\int_{-1}^{1} x^3\,dx}{\int_{-1}^{1} x^2\,dx} x - \frac{\int_{-1}^{1} x^2\,dx}{\int_{-1}^{1} dx} \cdot 1$$

$$= \tfrac{1}{3}(3x^2 - 1)$$

$$g_3 = f_3 - \frac{(f_3, g_2)}{(g_2, g_2)} g_2 - \frac{(f_3, g_1)}{(g_1, g_1)} g_1 - \frac{(f_3, g_0)}{(g_0, g_0)} g_0$$

$$= x^3 - \frac{\int_{-1}^{1} \tfrac{1}{3}(3x^2 - 1) x^2\,dx}{\int_{-1}^{1} \tfrac{1}{3}(3x^2 - 1)^2\,dx} \tfrac{1}{3}(3x^2 - 1) - \frac{\int_{-1}^{1} x^4\,dx}{\int_{-1}^{1} x^2\,dx} x$$

$$- \frac{\int_{-1}^{1} x^3\,dx}{\int_{-1}^{1} dx} 1$$

$$= \tfrac{1}{5}(5x^3 - 3x)$$

To normalize these elements, we note that

$$(g_0, g_0) = 2, \quad (g_1, g_1) = \tfrac{2}{3}, \quad (g_2, g_2) = \tfrac{8}{45},$$
$$(g_3, g_3) = \tfrac{8}{175}$$

and the required orthonormal basis is

$$\left\{ \frac{1}{\sqrt{2}}, \sqrt{\frac{3}{2}}x, \sqrt{\frac{5}{8}}(3x^2 - 1), \sqrt{\frac{7}{8}}(5x^3 - 3x) \right\}$$

These polynomials are the first four of the *Legendre polynomials* which are important in analysis. For further information on these and other applications of inner product spaces in analysis, e.g. Fourier series, see D.L. Kreider, R.G. Kuller, D.R. Ostberg and F.W. Perkins (*loc. cit.*).

Exercises 5.3

(Unless otherwise stated the inner products are the standard inner products.)

142

1. Show that each of the following pairs of vectors are orthogonal
    (i) $(2, 3, -2, 1, 0, 1)$ and $(2, -1, 1, 0, 2, 1)$ in $V_6(\mathbf{R})$
    (ii) $(i, 1, -i)$ and $(1 - i, 2, 1 + i)$ in $V_3(\mathbf{C})$
    (iii) 1 and $\cos \pi x$ in $C[0, 1]$.

2. Find all vectors which are orthogonal to the following
    (i) $(-1, 1, 2, -1)$ in $V_4(\mathbf{R})$
    (ii) $(1 - i, 1 + i)$ in $V_2(\mathbf{C})$

3. In $C[0, 1]$ prove that $\cos 2m\pi x$ and $\cos 2n\pi x$ $(m \neq n)$ are orthogonal and find a quadratic polynomial orthogonal to 1 and x. Furthermore, find the length of $\cos m\pi x$ and find a necessary and sufficient condition for $a + bx$ and $c + dx$ to be orthogonal in $C[0, 1]$.

4. Show that $\sin \pi x, \sin 2\pi x, \ldots, \sin n\pi x$ are orthogonal in $C[0, 1]$. Obtain an orthonormal set of functions from these.

5. Use the Gram-Schmidt Orthogonalization Procedure to orthogonalise
    (i) $\{(1, -1, 1), (2, 1, 1), (1, 0, 1)\}$ in $V_3(\mathbf{R})$
    (ii) $\{(1, -1, 1, 1), (0, 1, 0, 1), (2, -1, 1, 1)\}$ in $V_4(\mathbf{R})$
    (iii) $\{(1, -1, i), (i, 1, 2)\}$ in $V_3(\mathbf{C})$.

6. Complete to an orthonormal basis
    (i) $\left\{ \left(\frac{1}{\sqrt{2}}, 0, \frac{1}{\sqrt{2}} \right), (0, 1, 0) \right\}$ for $V_3(\mathbf{R})$
    (ii) $\{ \frac{1}{2}(1, i, 1, i), \frac{1}{2}(i, 1, i, 1)\}$ for $V_4(\mathbf{C})$.

7. Find orthonormal bases for $V_3(\mathbf{R})$ for the maps in Exercise 2 in §5.2 which are inner products.

8. Let V be the subspace of $C[0, 1]$ containing real polynomials of degree at most 3. Apply the Gram-Schmidt Orthogonalization Procedure to the $\mathbf{R}$-basis $\{1, x, x^2, x^3\}$ for V.

9. If $\{v_1, v_2, \ldots, v_n\}$ is an orthonormal basis for an inner product space V, prove that every $v \in V$ can be expressed as

$$v = \sum_{i=1}^{n} (v, v_i) v_i$$

    (i) Find an orthonormal basis for $V_3(\mathbf{C})$ and find the co-ordinates of $(1, i, -i)$ relative to this basis;
    (ii) Find an orthonormal basis for the subspace of $C[0, 1]$ consisting of polynomials of degree at most 2 and find the co-ordinates of $x^2 + 1$ and $x^2 - x + 1$ relative to this basis.

143

10. Apply the Gram-Schmidt Orthogonalization Procedure to the **R**-basis $\{1, x, x^2, x^3\}$ for $P_3(\mathbf{R})$ where the inner product is defined by

$$\int_{-1}^{1} (1 - x^2)\, f(x)\, g(x)\, dx$$

11. Let V be a finite dimensional inner product space with orthonormal basis $\mathscr{B} = \{v_1, \ldots, v_n\}$ and T a linear transformation on V. Prove that $(T)_{\mathscr{B}} = ((Tv_j, v_i))$.

12. If V is a Euclidean space and $u, v \in V$ are such that $\|u\| = \|v\|$, prove that $u - v$ is orthogonal to $u + v$. What does this say about the diagonals of a rhombus?

5.4 Application to the Rank of a Matrix

In order to apply the above to prove that the column rank of a matrix is equal to its row rank we need first to introduce some additional concepts. For the first part of this section, let V be an arbitrary K-space.

DEFINITION 5.8 *Let U and W be subspaces of V, then V is called the **direct sum** of U and W, written $V = U \oplus W$ if*
(i) $V = U + W$
(ii) *every element v of V can be **uniquely** expressed as $v = u + w$, where $u \in U$, $w \in W$.*

The first lemma gives an alternative criterion for V to be the direct sum of U and W.

LEMMA 5.9 *If $V = U + W$ then $V = U \oplus W$ if and only if $U \cap W = \{0\}$.*

PROOF If $v \in U \cap W$ then $v \in U$ and $v \in W$ then

$$v = v + 0 = 0 + v$$

and since $V = U \oplus W$, $v = 0$ and $U \cap W = \{0\}$.
Conversely, assume that $U \cap W = \{0\}$.
If $v \in W$, suppose that $v = u_1 + w_1 = u_2 + w_2$, where $u_1, u_2 \in U$, w_1, $w_2 \in W$ are two expressions for v, then $u_1 - u_2 = w_2 - w_1 \in U \cap W$ and so $u_1 = u_2$, $w_1 = w_2$ and the above expression for v is unique. ∎

LEMMA 5.10 *If V is a finite dimensional K-space and U is a subspace of V, then there exists a subspace W of V such that*

$$V = U \oplus W$$

and $(V:K) = (U:K) + (W:K)$

144

PROOF If $\{v_1, \ldots, v_m\}$ is a K-basis for U, extend this basis to give a K-basis $\{v_1, \ldots, v_m, v_{m+1}, \ldots, v_n\}$ for V. Let W be the subspace generated by $\{v_{m+1}, \ldots, v_n\}$, then it follows that $V = U \oplus W$ and $(V:K) = (U:K) + (W:K)$. ∎

From now on, let V be an inner product space.

DEFINITION 5.11 *If S is a subset of V, the **orthogonal complement** $S^\perp = \{x \in V | (x, s) = 0 \text{ for all } s \in S\}$.*

LEMMA 5.12 *If S is a subset of V, $S^\perp$ is a subspace of V.*

PROOF $S^\perp$ is non-empty since $0 \in S^\perp$.
If $u, v \in S^\perp, \alpha \in K$, then

$$(\alpha u + v, s) = \alpha(u, s) + (v, s) = 0 \quad \text{for all} \quad s \in S$$

and so $\alpha u + v \in S^\perp$ and $S^\perp$ is a subspace of V. ∎

THEOREM 5.13 *Let W be a subspace of a finite dimensional inner product space V, then*

$$V = W \oplus W^\perp$$

PROOF We can clearly assume that $W \neq \{0\}$ and $W \neq V$. If $(V:K) = n$ and $(W:K) = r < n$ then by Theorem 5.7 W has an orthonormal basis $\{w_1, w_2, \ldots, w_r\}$. Extend this to give a K-basis $\{w_1, \ldots, w_r, v_{r+1}, \ldots, v_n\}$ for V. Applying the Gram-Schmidt orthogonalization procedure to this K-basis will give an orthonormal basis $\{w_1, \ldots, w_r, \ldots, w_n\}$ for V. We show that $\{w_{r+1}, \ldots, w_n\}$ is a K-basis for $W^\perp$. If $v \in W^\perp$, then $v \in V$ and $v = \sum_{i=1}^{n} \alpha_i w_i$. If $1 \leqslant i \leqslant r$ then $0 = (v, w_i) = \left(\sum_{j=1}^{n} \alpha_j w_j, w_i\right) = \alpha_i$ and $v = \sum_{i=r+1}^{n} \alpha_i w_i$, i.e. $\{w_{r+1}, \ldots, w_n\}$ generates $W^\perp$.

Conversely, every K-linear combination of $\{w_{r+1}, \ldots, w_n\}$ is in $W^\perp$ and thus $\{w_{r+1}, \ldots, w_n\}$ is a K-basis for $W^\perp$, and by Lemma 5.9

$$V = W \oplus W^\perp$$

∎

We now apply this to the solution of a system of linear equations. Consider the system of linear equations

$$\sum_{j=1}^{n} \alpha_{ij} x_j = 0 \qquad (i = 1, 2, \ldots, m)$$

If $A = (\alpha_{ij}) \in M_{m,n}(K)$ and $x = (x_1, x_2, \ldots, x_n)$ then this system may be represented in matrix form as

$$Ax^t = 0$$

Now let $r_i = (\alpha_{i1}, \alpha_{i2}, \ldots, \alpha_{in}) \in V_n(K)$ $(i = 1, 2, \ldots, m)$, then this system can also be represented as

$$(r_i, x) = 0 \qquad (i = 1, 2, \ldots, m)$$

Let R_A be the subspace of $V_n(K)$ generated by $\{r_1, r_2, \ldots, r_m\}$. Thus the solution space of the system is $R_A^{\perp}$.
But, by Theorem 5.13

$$V_n(K) = R_A \oplus R_A^{\perp}$$

and the dimension of the solution space $= (R_A^{\perp} : K) = n - (R_A : K)$, where $(R_A : K)$ is the row rank of A. However, by Theorem 4.22, the dimension of the solution space is n-column rank of A. Thus, we have proved

THEOREM 5.14   *If $A \in M_{m,n}(K)$ then the row rank of A is equal to the column rank of A.*

We note that the above proof is valid only over the real and complex fields although the result is true for arbitrary fields as was seen in Chapter 4. The interested reader can look up, for example, S. Lang, Linear Algebra (Addision-Wesley), to see how the general case can be handled by similar methods, where (iv) of Definition 5.2 has been replaced by another condition (non-degeneracy of inner products) which allows the above to be applied to $V_n(K)$ for arbitrary K

CHAPTER 6

Diagonalization of Matrices and Linear Transformations

6.1 Introduction

Let $A, B \in M_n(K)$, then in Definition 4.5 of Chapter IV we have defined **similarity** of matrices by saying that B is similar to A (written $A \sim B$) if there exists an invertible matrix $P \in M_n(K)$ such that

$$B = P^{-1}AP$$

The following lemma is easily proved.

LEMMA 6.1   *Similarity of matrices is an equivalence relation on* $M_n(K)$.

PROOF $\sim$ is reflexive since $A = I^{-1}AI$ for all $A \in M_n(K)$. $\sim$ is symmetric since if $A \sim B$ there exists an invertible matrix $P \in M_n(K)$ such that $B = P^{-1}AP$ from which it follows that $A = (P^{-1})^{-1}BP^{-1}$ and $B \sim A$. Finally, $\sim$ is transitive since if $A \sim B$ and $B \sim C$ there exist invertible matrices $P, Q \in M_n(K)$ such that $B = P^{-1}AP$ and $C = Q^{-1}BQ$ and hence $C = Q^{-1}P^{-1}APQ = (PQ)^{-1}A(PQ)$, where PQ is an invertible matrix in $M_n(K)$. ∎

This means that $M_n(K)$ is partitioned into equivalence classes under this equivalence relation. We have the problem of determining representatives of these equivalence classes which are in as "simple" a form as possible. This will be one of the main goals of this chapter. This problem may also be formulated in terms of linear transformations.

Let V be a finite dimensional K-space with $(V : K) = n$ and $T \in \mathscr{L}(V)$. In Chapter IV, we saw that the matrices of T relative to different K-bases for V are similar to each other. Thus, the above problem is equivalent to that of determining a K-basis for V such that the matrix of T relative to that K-basis is as "simple" as possible. To this end, we define in the next section the eigenvalues and eigenvectors of a matrix (or of a linear transformation).

This may be further illustrated by considering an example. A linear transformation T on $V_2(\mathbf{R})$ is defined by

$$Te_1 = 2e_1 - e_2$$
$$Te_2 = 3e_1 - 2e_2$$

where $\{e_1, e_2\}$ is the standard basis. Given an arbitrary point $(\lambda, \mu) \in V_2(\mathbf{R})$, then

$$T(\lambda, \mu) = 2\lambda e_1 - \lambda e_2 + 3\mu e_1 - 2\mu e_2$$
$$= (2\lambda + 3\mu, -\lambda - 2\mu)$$

Now let $e_1' = 3e_1 - e_2$, $e_2' = e_1 - e_2$, then $\{e_1', e_2'\}$ is also a $\mathbf{R}$-basis for $V_2(\mathbf{R})$ and

$$Te_1' = 3(Te_1) - Te_2 = 6e_1 - 3e_2 - 3e_1 + 2e_2 = 3e_1 - e_2 = e_1'$$
$$Te_2' = Te_1 - Te_2 = 2e_1 - e_2 - 3e_1 + 2e_2 = -e_1 + e_2 = -e_2'$$

and $T(\lambda, \mu) = (\lambda, -\mu)$, where now (λ, μ) are the co-ordinates relative to the $\mathbf{R}$-basis $\{e_1', e_2'\}$. Thus the effect of the linear transformation T on an arbitrary point (or vector) is far clearer in terms of the second $\mathbf{R}$-basis, i.e. we have reflection in the first axis e'_1 as shown in the Fig. 1, i.e. to find the image of the point P we draw a line through P parallel to e_2', the image of P will be at a point P' so that $PX = XP'$, where X is the point where this line cuts the e_1'-axis.

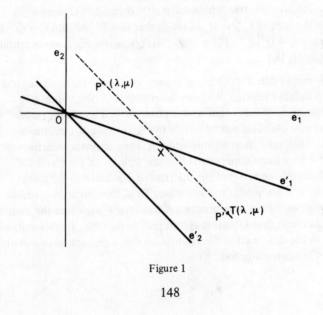

Figure 1

148

6.2 Eigenvalues and Eigenvectors

DEFINITION 6.2   (1) *If $A \in M_n(K)$, then an element $\lambda \in K$ is called an **eigenvalue** of A if there exists a non-zero $X \in V_n(K)$ such that $AX = \lambda X$ (X is a column vector). The vector X is called an **eigenvector** corresponding to the eigenvalue λ.* (2) *If V is an n-dimensional K-space and $T \in \mathscr{L}(V)$ then an element $\lambda \in K$ is called an **eigenvalue** of T, if there exists a non-zero $v \in V$ such that $Tv = \lambda v$, v is called an **eigenvector** corresponding to the eigenvalue λ.*

The connection between (1) and (2) is brought out by the following:

If A is the matrix of T relative to a K-basis $\mathscr{B} = \{v_1, \ldots, v_n\}$ and λ is an eigenvalue of T and v is an eigenvector corresponding to λ and

$$v = \sum_{i=1}^{n} \beta_i v_i \text{; put } X = \begin{pmatrix} \beta_1 \\ \beta_2 \\ . \\ . \\ . \\ \beta_n \end{pmatrix}, \text{ then } AX = \lambda X, \text{ that is, } \lambda \text{ is an eigenvalue}$$

of A. Furthermore, if $\mathscr{B}' = \{v_1', \ldots, v_n'\}$ is another K-basis for V, then

$(T)_{\mathscr{B}'} = PAP^{-1}$, where $P = (p_{ij})$ and $v_j = \sum_{i=1}^{n} p_{ij} v_i'$ $(j = 1, \ldots, n)$

(see p.98) and hence

$$(PAP^{-1})(PX) = \lambda(PX)$$

i.e. λ is also an eigenvalue of PAP^{-1} with corresponding eigenvector PX (which is non-zero since P is invertible). Thus, we could have defined the eigenvalues of a linear transformation to be the eigenvalue of the matrix of T relative to any K-basis for V. In the sequel, we concentrate on matrices but will state and prove some of the results in terms of linear transformations.

If $\lambda \in K$ is an eigenvalue of $A \in M_n(K)$ then from Definition 6.2 there exists a non-zero column vector X such that

$$(A - \lambda I)X = 0$$

By Corollary 3 to Theorem 2.7 such an X exists if and only if $\det(A - \lambda I) = 0$.

Thus we have proved

THEOREM 6.3 $\lambda \in K$ *is an eigenvalue of A if and only if $\det(A - \lambda I) = 0$.*

If $A \in M_n(K)$, then $\det(A - xI)$ is a polynomial of degree n in x.

DEFINITION 6.4 $\chi_A(x) = det\,(A - xI)$ is called the **characteristic polynomial** of the matrix A.

Thus, Theorem 6.3 may alternatively be stated as

THEOREM 6.3′ $\lambda \in K$ is an eigenvalue of A if and only if it is a root of the characteristic polynomial of A.

This gives us a practical method for computing the eigenvalues of A. $\chi_A(x)$ is a polynomial of degree n in x; by the so-called fundamental theorem of algebra, every polynomial of degree n over the complex field factorises completely into n linear factors over $\mathbf{C}$, i.e. if $\chi_A(x) = (-1)^n(x^n + a_1 x^{n-1} + \ldots + a_n)$, where $a_1, a_2, \ldots, a_n \in \mathbf{C}$, then

$$\chi_A(x) = (-1)^n (x - \lambda_1)(x - \lambda_2) \ldots (x - \lambda_n)$$

where $\lambda_1, \lambda_2, \ldots, \lambda_n \in \mathbf{C}$. Thus over the complex field, an $n \times n$ matrix has at most n eigenvalues (and has at least one eigenvalue). It is also clear that the existence of eigenvalues depends on the field over which we are working.

EXAMPLES

1. If $A = \begin{pmatrix} 0 & 1 \\ -1 & 0 \end{pmatrix}$ then

$$\chi_A(x) = \begin{vmatrix} -x & 1 \\ -1 & -x \end{vmatrix} = x^2 + 1$$

which is irreducible over $\mathbf{R}$ (i.e. does not factorise in $\mathbf{R}$), but it factorises over $\mathbf{C}$ to give

$$\chi_A(x) = (x + i)(x - i)$$

Hence A has **no** eigenvalues in $\mathbf{R}$ but has the eigenvalues i and $-$i in $\mathbf{C}$.

2. If $A = \begin{pmatrix} 2 & 1 \\ -1 & 0 \end{pmatrix}$ then

$$\chi_A(x) = \begin{vmatrix} 2-x & 1 \\ -1 & -x \end{vmatrix} = (1-x)\begin{vmatrix} 1 & 1 \\ -1 & -x \end{vmatrix} = (x-1)^2$$

Thus A has **one** eigenvalue $\lambda = 1$ in $\mathbf{R}$ (and also in $\mathbf{C}$).

Once the eigenvalues have been determined, the corresponding eigenvectors are easily calculated, i.e. if $A = (\alpha_{ij})$ and λ is an eigenvalue of A and $X = (x_1, x_2, \ldots, x_n)$ is the corresponding eigenvector, then

150

from Definition 6.2 X is a non-trivial solution of the system of linear equations

$$
\left.
\begin{aligned}
(\alpha_{11} - \lambda)x_1 + \alpha_{12}x_2 + \ldots + \alpha_{1n}x_n &= 0 \\
\alpha_{21}x_1 + (\alpha_{22} - \lambda)x_2 + \ldots + \alpha_{2n}x_n &= 0 \\
&\vdots \\
\alpha_{n1}x_1 + \alpha_{n2}x_2 + \ldots + (\alpha_{nn} - \lambda)x_n &= 0
\end{aligned}
\right\}
$$

A non-trivial solution certainly exists since $\det(A - \lambda I) = 0$. We shall see later (Theorem 6.9), that the number of linearly independent non-trivial solutions is no higher than the multiplicity of $x - \lambda$, in $\chi_A(x)$.

In the above two examples, if $A = \begin{pmatrix} 0 & 1 \\ -1 & 0 \end{pmatrix}$, corresponding to the eigenvalue i, we must solve

$$
\left.
\begin{aligned}
-ix_1 + x_2 &= 0 \\
-x_1 - ix_2 &= 0
\end{aligned}
\right\},
$$

Since reduction of the matrix of coefficients gives

$$
\begin{pmatrix} -i & 1 \\ -1 & -i \end{pmatrix} \to \begin{pmatrix} -i & 1 \\ 0 & 0 \end{pmatrix}
$$

then all solutions are of the form $\alpha(1, i)$, $\alpha \in \mathbf{C}$ and thus $\begin{pmatrix} 1 \\ i \end{pmatrix}$ is an eigenvector corresponding to the eigenvalue $\lambda = i$, that is

$$
\begin{pmatrix} -i & 1 \\ -1 & -i \end{pmatrix} \begin{pmatrix} 1 \\ i \end{pmatrix} = \begin{pmatrix} 0 \\ 0 \end{pmatrix}.
$$

Similarly, if $\lambda = -i$ then

$$
\begin{pmatrix} i & 1 \\ -1 & i \end{pmatrix} \begin{pmatrix} i \\ 1 \end{pmatrix} = \begin{pmatrix} 0 \\ 0 \end{pmatrix}
$$

and $\begin{pmatrix} i \\ 1 \end{pmatrix}$ is an eigenvector corresponding to the eigenvalue $\lambda = -i$.

In the second example, when $A = \begin{pmatrix} 2 & 1 \\ -1 & 0 \end{pmatrix}$ then $\chi_A(\lambda) = (\lambda - 1)^2$ and

$$
\begin{pmatrix} 1 & 1 \\ -1 & -1 \end{pmatrix} \begin{pmatrix} 1 \\ -1 \end{pmatrix} = \begin{pmatrix} 0 \\ 0 \end{pmatrix}
$$

and there is only **one** linearly independent eigenvector corresponding to $\lambda = 1$.

EXAMPLE   As a further example we have

$$A = \begin{pmatrix} 3 & 2 & 4 \\ 2 & 0 & 2 \\ 4 & 2 & 3 \end{pmatrix}$$

then the characteristic polynomial $\chi_A(x)$ is given by

$$\chi_A(x) = \begin{vmatrix} 3-x & 2 & 4 \\ 2 & -x & 2 \\ 4 & 2 & 3-x \end{vmatrix} = \begin{vmatrix} -1-x & 2 & 4 \\ 0 & -x & 2 \\ 1+x & 2 & 3-x \end{vmatrix}$$

$$= (1+x) \begin{vmatrix} -1 & 2 & 4 \\ 0 & -x & 2 \\ 1 & 2 & 3-x \end{vmatrix} = (x+1) \begin{vmatrix} -x & 2 \\ 4 & 7-x \end{vmatrix}$$

$$= -(x+1)^2(x-8)$$

The eigenvalues over $\mathbf{R}$ are -1 and 8.

Corresponding to the eigenvalue $\lambda = -1$, we have the system of linear equations

$$\left. \begin{array}{r} 4x_1 + 2x_2 + 4x_3 = 0 \\ 2x_1 + x_2 + 2x_3 = 0 \\ 4x_1 + 2x_2 + 4x_3 = 0 \end{array} \right\},$$

which clearly reduce to the one equation.

$$2x_1 + x_2 + 2x_3 = 0$$

i.e. $\quad x_2 = -2x_1 - 2x_3$

and $(1, -2, 0)$ and $(0, -2, 1)$ are two linearly independent solutions. Thus, corresponding to the eigenvalue $\lambda = -1$, there are **two** linearly independent eigenvectors.

If $\lambda = 8$, then $\begin{pmatrix} -5 & 2 & 4 \\ 2 & -8 & 2 \\ 4 & 2 & -5 \end{pmatrix} \begin{pmatrix} 2 \\ 1 \\ 2 \end{pmatrix} = \begin{pmatrix} 0 \\ 0 \\ 0 \end{pmatrix}$ and $\begin{pmatrix} 2 \\ 1 \\ 2 \end{pmatrix}$ is an

eigenvector corresponding to the eigenvalue $\lambda = 8$.

In order that the above ideas may be extended to cover linear transformations T, we prove the following lemma.

LEMMA 6.5 *Similar matrices have the same characteristic polynomial and hence the same eigenvalues.*

152

PROOF   If A and $B \in M_n(K)$ are similar matrices, there exists an invertible matrix P such that $B = PAP^{-1}$. Now the characteristic polynomial of B is

$$\begin{aligned}
\chi_B(x) &= \det(B - xI_n) = \det(PAP^{-1} - xI_n) \\
&= \det(P(A - xI_n)P^{-1}) \\
&= (\det P)\det(A - xI_n)\det(P^{-1}) \\
&= \det(A - xI_n) \\
&= \chi_A(x)
\end{aligned}$$

where we have used the fact that the determinant function is multiplicative (Corollary 2 to Theorem 2.7) and that $\det(P^{-1}) = (\det P)^{-1}$. ∎

Thus, if $T \in \mathscr{L}(V)$, since the matrices of T relative to distinct K-bases for V are similar to each other, (see Theorem 4.6) the following definition of the characteristic polynomial of T is unambiguous.

DEFINITION 6.6   *If $T \in \mathscr{L}(V)$, the **characteristic polynomial** of T is the characteristic polynomial of the matrix of T relative to any K-basis for V.*

We can now prove

LEMMA 6.7   *Let λ be an eigenvalue of $A \in M_n(K)$ $(T \in \mathscr{L}(V))$ and let $V(\lambda)$ denote the set of eigenvectors corresponding to λ together with the zero vector. Then $V(\lambda)$ is a subspace of $V_n(K)(V)$.*

PROOF   It is clear that $V(\lambda)$ is the solution space of the system of linear equations represented by

$$(A - \lambda I_n)X = 0$$

and by §3.3, Example 4, it is a subspace of $V_n(K)$. (Similarly $V(\lambda) = \ker(T - \lambda I_V)$ which by Lemma 4.7 is a subspace of V). ∎

DEFINITION 6.8   *$V(\lambda)$ is called the **eigenspace** corresponding to the eigenvalue λ.*

The next theorem gives an upper limit on the dimension of $V(\lambda)$, that is, on the number of linearly independent eigenvectors corresponding to the eigenvalue λ.

THEOREM 6.9   *If λ is an eigenvalue of $A \in M_n(K)$ $(T \in \mathscr{L}(V))$, then $(V(\lambda):K) \leqslant$ multiplicity of $(x - \lambda)$ as a factor in the characteristic polynomial $\chi_A(x)$ $(\chi_T(x))$.*

153

PROOF   We shall give the proof in terms of linear transformations.

Suppose that $(V(\lambda):K) = r$, let $\{v_1, \ldots, v_r\}$ be a K-basis for $V(\lambda)$, which is extended to give a K-basis $\mathscr{B} = \{v_1, \ldots, v_r, \ldots, v_n\}$ for V. Then we have $Tv_j = \lambda v_j \ (j = 1, \ldots, r)$ and

$$(T)_{\mathscr{B}} = r\left\{ \left(\begin{array}{ccc|c} \lambda & \cdots & 0 & \\ \cdot & & \cdot & \\ \cdot & & \cdot & * \\ \cdot & & \cdot & \\ 0 & \cdots & \lambda & \\ \hline & 0 & & A \end{array} \right) \right.$$

where we are not interested in the explicit values in the part indicated $*$ and A is an $(n - r) \times (n - r)$ matrix. Thus

$$\chi_T(x) = \det((T)_{\mathscr{B}} - xI_n)$$

$$= \det \left(\begin{array}{ccc|c} \lambda - x & \cdots & 0 & \\ \vdots & & \vdots & * \\ 0 & \cdots & \lambda - x & \\ \hline & 0 & & A - xI_{n-r} \end{array} \right)$$

$$= (\lambda - x)^r \det(A - xI_{n-r})$$

Hence $(\lambda - x)^r$ is a factor of the characteristic polynomial of T and thus $r \leqslant$ multiplicity of $(x - \lambda)$ as a factor in the characteristic polynomial $\chi_T(x)$ of T. ∎

Exercises 6.2

1. Find the eigenvalues of the following matrices over (a) the rational field **Q**, (b) the real field **R**, (c) the complex field **C**.

(i) $\begin{pmatrix} 1 & \sqrt{2} \\ -\sqrt{2} & 1 \end{pmatrix}$ (ii) $\begin{pmatrix} 1 & -1 & -1 \\ 1 & -1 & 0 \\ 1 & 0 & -1 \end{pmatrix}$

(iii) $\begin{pmatrix} 1 & 0 & 0 & 0 & 0 \\ 0 & 1 & 1 & 0 & 0 \\ 0 & 1 & -1 & 0 & 0 \\ 0 & 0 & 0 & 1 & 2 \\ 0 & 0 & 0 & -1 & -1 \end{pmatrix}$

154

2. Find the characteristic polynomial, eigenvalues and eigenvectors of the following matrices over the complex field

(i) $\begin{pmatrix} 1 & 0 & -1 \\ 1 & 2 & 1 \\ 2 & 2 & 3 \end{pmatrix}$ (ii) $\begin{pmatrix} 0 & 1 & 0 \\ 0 & 0 & 1 \\ 1 & -3 & 3 \end{pmatrix}$

(iii) $\begin{pmatrix} 2 & i & 1+2i \\ -i & 0 & -i \\ 1-2i & i & 0 \end{pmatrix}$ (iv) $\begin{pmatrix} 1 & 1 & 1 & 0 \\ 1 & 1 & 0 & 1 \\ 1 & 0 & 1 & 1 \\ 0 & 1 & 1 & 1 \end{pmatrix}$

(v) $\begin{pmatrix} 3 & 2 & 2 & -4 \\ 2 & 3 & 2 & -1 \\ 1 & 1 & 2 & -1 \\ 2 & 2 & 2 & -1 \end{pmatrix}$ (vi) $\begin{pmatrix} 1 & -1 & 1 & 1 & 0 \\ 0 & 0 & 2 & 2 & 2 \\ 0 & 1 & 0 & -1 & -1 \\ -2 & 0 & -2 & -3 & -2 \\ 2 & 0 & 2 & 4 & 3 \end{pmatrix}$

3. If $A, B \in M_n(K)$, prove that AB and BA have the same eigenvalues.

4. If $\lambda_1, \ldots, \lambda_n$ are the eigenvalues of $A \in M_n(K)$, prove that
   (i) if A is invertible, $1/\lambda_1, \ldots, 1/\lambda_n$ are the eigenvalues of A^{-1},
   (ii) $\lambda_1^k, \ldots, \lambda_n^k$ are the eigenvalues of A^k $(k = 1, 2, \ldots)$.
What are the corresponding eigenvectors?

5. Find the eigenvalues and eigenvectors of the differentiation transformation D on $P_n(\mathbf{R})$.

6. If $A \in M_n(K)$, prove that A and A^t have the same eigenvalues.

7. Find the eigenvalues and eigenvectors of reflections and rotations in $V_2(\mathbf{R})$ over the real field.

8. Find the characteristic polynomial of the following $n \times n$ matrices

(i) $\begin{pmatrix} 0 & 0 & \ldots & 0 & \alpha_1 \\ 1 & 0 & \ldots & 0 & \alpha_2 \\ 0 & 1 & \ldots & 0 & \alpha_3 \\ \cdot & \cdot & & \cdot & \cdot \\ \cdot & \cdot & & \cdot & \cdot \\ 0 & 0 & \ldots & 1 & \alpha_n \end{pmatrix}$ (ii) $\begin{pmatrix} 1+b & a & a^2 & \ldots & a^{n-1} \\ 1 & a+b & a^2 & \ldots & a^{n-1} \\ 1 & a & a^2+b & \ldots & a^{n-1} \\ \cdot & \cdot & \cdot & & \cdot \\ \cdot & \cdot & \cdot & & \cdot \\ 1 & a & a^2 & \ldots & a^{n-1}+b \end{pmatrix}$

6.3 Diagonalization of Matrices

DEFINITION 6.10 (i) *A matrix $A \in M_n(K)$ is **diagonalizable** if there exists an invertible matrix P such that $P^{-1}AP$ is a diagonal matrix.*

(ii) *A linear transformation $T \in \mathscr{L}(V)$ is **diagonalizable** if there exists a K-basis for V such that the matrix of T relative to this K-basis is a diagonal matrix.*

It is clear from what has been said earlier that these two definitions are equivalent.

We now prove a necessary and sufficient condition for a matrix to be diagonalizable.

THEOREM 6.11 *A matrix $A \in M_n(K)$ $(T \in \mathscr{L}(V))$ is diagonalizable if and only if a set of eigenvectors of $A(T)$ form a K-basis for $V_n(K)(V)$.*

PROOF If A is diagonalizable, then by Definition 6.10 there exists an invertible matrix P such that $P^{-1}AP = D = \mathrm{diag}\,(\lambda_1, \lambda_2, \ldots, \lambda_n)$. Since similar matrices have the same eigenvalues, the eigenvalues of A are $\lambda_1, \ldots, \lambda_n$. Let $P = (C_1, C_2, \ldots, C_n)$, where C_i $(i = 1, \ldots, n)$ indicate the columns of P. Since P is invertible, $\{C_1, \ldots, C_n\}$ is linearly independent over K and thus forms a K-basis for $V_n(K)$. Now

$$AP = PD$$

or $A(C_1, C_2, \ldots, C_n) = (C_1, C_2, \ldots, C_n)\,\mathrm{diag}\,(\lambda_1, \ldots, \lambda_n)$ implies that

$$AC_i = \lambda_i C_i \qquad (i = 1, \ldots, n)$$

that is C_i is an eigenvector corresponding to the eigenvalue λ_i of A.

Let $\{X_1, X_2, \ldots, X_n\}$ be a linearly independent set of eigenvectors of A. Suppose that these eigenvectors correspond to the eigenvalues $\lambda_1, \lambda_2, \ldots, \lambda_n$ (not necessarily distinct). If $\{X_1, X_2, \ldots, X_n\}$ is linearly independent over K then if $P = (X_1, X_2, \ldots, X_n)$ is the $n \times n$ matrix formed with X_i $(i = 1, \ldots, n)$ as its columns then by the Corollary to Theorem 4.18, P is invertible and by inverting the above argument we have that

$$AX_i = \lambda_i X_i \qquad (i = 1, \ldots, n)$$

implies that

$$P^{-1}AP = \mathrm{diag}\,(\lambda_1, \lambda_2, \ldots, \lambda_n)$$

∎

The above theorem implies that if A is diagonalizable then $\chi_A(x)$ factors completely into linear factors.

We can now prove that

THEOREM 6.12 *If* $A \in M_n(K)$ *($T \in \mathscr{L}(V)$) has n distinct eigenvalues then A(T) is diagonalizable.*

PROOF Let $\lambda_1, \lambda_2, \ldots, \lambda_n$ be the *n* distinct eigenvalues of *A* and $X_1, X_2, \ldots, X_n$ be the corresponding eigenvectors, thus $AX_i = \lambda_i X_i$ ($i = 1, \ldots, n$). We need only prove that $\{X_1, X_2, \ldots, X_n\}$ is linearly independent over *K* then the proof will be complete by Theorem 6.11. The proof of this is by induction on *n*. This is clearly true when $n = 1$. We shall assume that $\{X_1, X_2, \ldots, X_{r-1}\}$ is linearly independent over *K*, where $1 \leqslant r - 1 < n$. Consider

$$\alpha_1 X_1 + \alpha_2 X_2 + \ldots + \alpha_r X_r = 0$$

where $\alpha_i \in K$ ($i = 1, \ldots, r$).

Premultiplying by *A* gives

$$\alpha_1 \lambda_1 X_1 + \alpha_2 \lambda_2 X_2 + \ldots + \alpha_r \lambda_r X_r = 0$$

and subtracting this from λ_r times the previous equation gives

$$\alpha_1 (\lambda_r - \lambda_1) X_1 + \ldots + \alpha_{r-1} (\lambda_r - \lambda_{r-1}) X_{r-1} = 0$$

But $\{X_1, \ldots, X_{r-1}\}$ is linearly independent over *K* and so

$$\alpha_i (\lambda_r - \lambda_i) = 0 \qquad (i = 1, \ldots, r-1)$$

Furthermore $(\lambda_r - \lambda_i) \neq 0$ ($i = 1, \ldots, r-1$) since the eigenvalues are distinct and so $\alpha_1 = \alpha_2 = \ldots = \alpha_{r-1} = 0$ and $\alpha_r X_r = 0$, and since $X_r \neq 0, \alpha_r = 0$, i.e. $\{X_1, \ldots, X_r\}$ is linearly independent over *K* $\{1 \leqslant r \leqslant n\}$. ∎

From the proof of Theorems 6.9 and 6.11 we have the following

COROLLARY *If* $A \in M_n(K)$ *is diagonalizable then* $(V(\lambda) : K) = $ *multiplicity of* $(x - \lambda)$ *as a factor in* $\chi_A(x)$ *for each eigenvalue* λ *of A.*

We note that the converse is also true when the characteristic polynomial factors completely into linear factors, e.g. when $K = \mathbf{C}$. We refer back to the earlier examples (pp. 151–152).

EXAMPLES

1. If $A = \begin{pmatrix} 0 & 1 \\ -1 & 0 \end{pmatrix}$, put $P = \begin{pmatrix} 1 & i \\ i & 1 \end{pmatrix}$, then

$$P^{-1} A P = \operatorname{diag}(i, -i)$$

2. If $A = \begin{pmatrix} 2 & 1 \\ -1 & 0 \end{pmatrix}$, then since $(1, -1)$ is the only linearly independent eigenvector corresponding to the only eigenvalue 1, $(V(1) : \mathbf{R}) = 1$ and by Theorem 6.11, A is not diagonalizable.

3. If $A = \begin{pmatrix} 3 & 2 & 4 \\ 2 & 0 & 2 \\ 4 & 2 & 3 \end{pmatrix}$, the eigenvalues are $-1, 8$.

We have shown that $\{(1, -2, 0), (0, -2, 1)\}$ is a $\mathbf{R}$-basis for $V(-1)$ and $\{(2, 1, 2)\}$ is a $\mathbf{R}$-basis for $V(8)$. If we put

$$P = \begin{pmatrix} 1 & 0 & 2 \\ -2 & -2 & 1 \\ 0 & 1 & 2 \end{pmatrix}$$

then

$$P^{-1}AP = \operatorname{diag}(-1, -1, 8)$$

We now assess the progress which has been made on the problem which was the main motivation for the work in this chapter, namely that of finding matrices of a "simple" form to be representatives of the equivalence classes under the equivalence relation of similarity. We have seen that in certain circumstances the "simple" form chosen is a diagonal matrix, but unfortunately, as we have seen in Example 2 above this simple form cannot always be chosen to be a diagonal matrix. We now state the solution in the general case without proof, the proof being beyond the scope of this book.

We assume that $K = \mathbf{C}$ the complex field.

Let $A \in M_n(K)$ have characteristic polynomial

$$\chi_A(x) = (-1)^n (x - \lambda_1)^{n_1}(x - \lambda_2)^{n_2} \ldots (x - \lambda_s)^{n_s}$$

where $\lambda_1, \lambda_2, \ldots, \lambda_s$ are the s distinct eigenvalues of A and $n_1 + n_2 + \ldots + n_s = n$.

Let

$$J(\lambda, r) = \begin{pmatrix} \lambda & 1 & 0 & 0 & \ldots & 0 \\ 0 & \lambda & 1 & 0 & \ldots & 0 \\ 0 & 0 & \ddots & & \ddots & \\ & & & \ddots & & 0 \\ \vdots & \vdots & & & \ddots & 1 \\ 0 & 0 & \ldots & & & \lambda \end{pmatrix} \in M_r(K)$$

158

If now $m_{i1}, m_{i2}, \ldots, m_{is_i}$ are positive integers such that $m_{i1} \geqslant m_{i2} \geqslant \ldots \geqslant m_{is_i}$ and

$$m_{i1} + m_{i2} + \ldots + m_{is_i} = n_i$$

let

$$J(n_i) = \begin{pmatrix} J(\lambda_i, m_{i1}) & 0 & \cdots & & 0 \\ 0 & J(\lambda_i, m_{i2}) & & & \\ \cdot & & & & \\ \cdot & & & \cdot & 0 \\ 0 & & & & J(\lambda_i, m_{is_i}) \end{pmatrix}$$

then the matrix A is similar to the matrix

$$J = \begin{pmatrix} J(n_1) & 0 & \cdots & 0 \\ 0 & J(n_2) & & \\ \cdot & & \cdot & \\ \cdot & & & \\ 0 & & & J(n_s) \end{pmatrix}$$

for some choice of m_{ij} $(i = 1, \ldots, s; j = 1, 2, \ldots, s_i)$ satisfying the above condition. Such a J is called the **Jordan Canonical Form of** A. It will be noted that J has the eigenvalues of A as diagonal elements and a distribution of zeros and ones on the superdiagonal and zeros elsewhere in the matrix.

This is best illustrated by an example, in the case $n = 3$, every 3×3 matrix A with complex elements will be similar to one of the following

$$\begin{pmatrix} \lambda & 1 & 0 \\ 0 & \lambda & 1 \\ 0 & 0 & \lambda \end{pmatrix}, \begin{pmatrix} \lambda & 1 & 0 \\ 0 & \lambda & 0 \\ 0 & 0 & \lambda \end{pmatrix}, \begin{pmatrix} \lambda & 0 & 0 \\ 0 & \lambda & 0 \\ 0 & 0 & \lambda \end{pmatrix}, \begin{pmatrix} \lambda & 1 & 0 \\ 0 & \lambda & 0 \\ 0 & 0 & \mu \end{pmatrix},$$

$$\begin{pmatrix} \lambda & 0 & 0 \\ 0 & \lambda & 0 \\ 0 & 0 & \mu \end{pmatrix}, \begin{pmatrix} \lambda & 0 & 0 \\ 0 & \mu & 0 \\ 0 & 0 & \nu \end{pmatrix}$$

Exercises 6.3

1. When possible, find an invertible matrix P such that $P^{-1}AP$ is a diagonal matrix if A is

$$\text{(i)} \begin{pmatrix} 1 & -1 & -1 \\ 1 & -1 & 0 \\ 1 & 0 & -1 \end{pmatrix} \quad \text{(ii)} \begin{pmatrix} 1 & 0 & -1 \\ 1 & 2 & 1 \\ 2 & 2 & 3 \end{pmatrix} \quad \text{(iii)} \begin{pmatrix} -2 & -8 & -12 \\ 1 & 4 & 4 \\ 0 & 0 & 1 \end{pmatrix}$$

$$\text{(iv)} \begin{pmatrix} 2-i & 0 & i \\ 0 & 1+i & 0 \\ i & 0 & 2-i \end{pmatrix} \quad \text{(v)} \begin{pmatrix} -1 & -1 & -6 & 3 \\ 1 & -2 & -3 & 0 \\ -1 & 1 & 0 & 1 \\ -1 & -1 & -5 & 3 \end{pmatrix}$$

$$\text{(vi)} \begin{pmatrix} 3 & -1 & -2 & 2 \\ 3 & -1 & -2 & 2 \\ 2 & -1 & -1 & 2 \\ 1 & 0 & 0 & 1 \end{pmatrix} \quad \text{(vii)} \begin{pmatrix} 1 & 1 & 1 & 1 \\ 0 & 1 & 1 & 1 \\ 0 & 0 & -1 & -1 \\ 0 & 0 & 0 & -1 \end{pmatrix}$$

2. Find the eigenvalues and eigenvectors of each of the following matrices

$$A = \begin{pmatrix} 6 & -3 \\ 5 & -2 \end{pmatrix}, \quad B = \begin{pmatrix} 3 & 1 \\ -1 & 1 \end{pmatrix}$$

over $\mathbf{Q}$. Show that A is similar to a diagonal matrix and explain why B is not. Find an invertible matrix P such that $P^{-1}AP$ is a diagonal matrix.

3. Determine the eigenvalues of the matrix

$$A = \begin{pmatrix} 1 & a & a^2 & 0 \\ 1 & a & 0 & a^2 \\ 1 & 0 & a & a^2 \\ 0 & 1 & a & a^2 \end{pmatrix}$$

Find an invertible matrix P such that $P^{-1}AP$ is a diagonal matrix.

4. If $A = \begin{pmatrix} 1 & 1 & 2 \\ 0 & 2 & 1 \\ 0 & 0 & 3 \end{pmatrix}$ and P is an invertible matrix, by considering $(P^{-1}AP)^n$ or otherwise, find A^n, where n is a positive integer.

5. Find an invertible matrix P, such that $P^{-1}AP$ is a diagonal matrix if

$$A = \begin{pmatrix} -1 & 4 & -2 \\ 1 & -1 & 1 \\ 3 & -6 & 4 \end{pmatrix}$$

Hence, find a matrix $X (\neq \pm A)$ satisfying $X^2 = A$.

6.4 The Minimum Polynomial of a Matrix and the Cayley-Hamilton Theorem

Throughout this section we assume that K is the complex field.

In §3.4, we saw that $(M_n(K) : K) = n^2$. Thus, if $A \in M_n(K)$, the set $\{I, A, A^2, \ldots, A^{n^2}\}$ is linearly dependent over K, i.e. there exist $\alpha_0, \alpha_1, \ldots, \alpha_{n^2} \in K$, not all zero, such that

$$\alpha_0 I + \alpha_1 A + \ldots + \alpha_{n^2} A^{n^2} = 0$$

If we put

$$f(x) = \alpha_0 + \alpha_1 x + \ldots + \alpha_{n^2} x^{n^2}$$

then

$$f(A) = 0$$

If now m is a positive integer such that $\{I, A, A^2, \ldots, A^{m-1}\}$ is linearly independent over K and $\{I, A, A^2, \ldots, A^m\}$ is linearly dependent over K, then there exist $\beta_0, \beta_1, \ldots, \beta_m \in K$ with $\beta_m \neq 0$ such that

$$\beta_0 I + \beta_1 A + \ldots + \beta_m A^m = 0$$

or

$$\gamma_0 I + \gamma_1 A + \ldots + \gamma_{m-1} A^{m-1} + A^m = 0$$

where $\gamma_i = \beta_i / \beta_m$ $(i = 0, 1, \ldots, m - 1)$. Hence, there exists a **monic** polynomial $f(x)$ of degree $m \leqslant n^2$ such that $f(A) = 0$. We can now give the following definition:

DEFINITION 6.13   *A **minimum polynomial** of a matrix $A \in M_n(K)$ is the monic polynomial $m(x)$ of least degree such that $m(A) = 0$.*

The above argument shows that such a polynomial exists and is of degree at most n^2. Our aim will be to obtain further restrictions on the degree and form of minimum polynomial which will prove useful in its evaluation.

The minimum polynomial of a linear transformation $T \in \mathscr{L}(V)$ can be defined in a similar way by using precisely the same arguments in the vector space $\mathscr{L}(V)$.

We now prove a series of useful lemmas concerning the minimum polynomial.

LEMMA 6.14   *The minimum polynomial of $A \in M_n(K)$ is unique.*

PROOF   Let $m(x)$ and $m'(x)$ be minimum polynomials of the matrix A. They are clearly of the same degree. Let $f(x) = m(x) - m'(x)$, then $f(x)$ is of lower degree than $m(x)$ and $m'(x)$ (since both are monic) and $f(A) = m(A) - m'(A) = 0$, which leads to a monic polynomial of lower degree than $m(x)$ satisfying the requirements for a minimum polynomial and which contradicts the definition of $m(x)$. Thus $m(x) - m'(x) = 0$, or $m(x) = m'(x)$. ∎

LEMMA 6.15   *The minimum polynomial $m(x)$ of $A \in M_n(K)$ divides every polynomial $f(x)$ such that $f(A) = 0$.*

PROOF   By the division algorithm for polynomials, there exist polynomials $q(x), r(x)$ such that

$$f(x) = q(x) m(x) + r(x)$$

where $r(x) = 0$ or the degree $r(x) <$ degree $m(x)$. Now

$$r(A) = f(A) - q(A) m(A) = 0$$

which means that if degree $r(x) \geqslant 1$, we have a polynomial of lower degree than $m(x)$ satisfying the requirements for a minimum polynomial. Thus $r(x) = 0$ and $m(x)$ divides $f(x)$. ∎

LEMMA 6.16   *Similar matrices have the same minimum polynomial.*

PROOF   If $A, B \in M_n(K)$ are similar, then there exists an invertible matrix P such that

$$P^{-1}AP = B$$

It is easily shown by induction that for $k \geqslant 1$

$$P^{-1}A^k P = B^k$$

and if $f(x) = \alpha_0 + \alpha_1 x + \ldots + \alpha_m x^m$ then

$$
\begin{aligned}
f(B) &= \alpha_0 I + \alpha_1 B + \ldots + \alpha_m B^m \\
&= \alpha_0 P^{-1}P + \alpha_1 P^{-1}AP + \ldots + \alpha_m P^{-1}A^m P \\
&= P^{-1}f(A) P
\end{aligned}
$$

Thus if $f(A) = 0$ then $f(B) = 0$ and conversely.
The result now follows from Lemma 6.15. ∎

We now prove an important theorem which gives further restrictions on the degree of the minimum polynomial, indeed that the degree of $m(x) \leqslant n$, if $A \in M_n(K)$.

THEOREM 6.17 *(Cayley-Hamilton Theorem) If $\chi_A(x)$ is the characteristic polynomial of A, then $\chi_A(A) = 0$ or alternatively, "every matrix satisfies its characteristic polynomial".*

PROOF If $A \in M_n(K)$, let the characteristic polynomial of A be

$$\chi_A(x) = (-1)^n (x^n + a_1 x^{n-1} + \ldots + a_n)$$

By §2.4, we have

$$(A - xI)\, \text{adj}\,(A - xI) = |A - xI|\,.\,I = \chi_A(x)\,.\,I$$

Now, by considering the form of $\text{adj}(A - xI)$, we have
$\text{adj}(A - xI) = (p_{ij}(x))$, where $p_{ij}(x)$ are polynomials of degree at most $n - 1$ in x. Thus $\text{adj}(A - xI)$ may be expressed in the form

$$\text{adj}(A - xI) = B_0 + B_1 x + \ldots + B_{n-1} x^{n-1}$$

where $B_0, B_1, \ldots, B_{n-1} \in M_n(K)$. By comparing the coefficients of powers of x in

$$(A - xI)(B_0 + B_1 x + \ldots + B_{n-1} x^{n-1}) = (-1)^n (x^n + a_1 x^{n-1} + \ldots + a_n)I$$

we obtain

$$AB_0 = (-1)^n a_n I$$
$$-B_0 + AB_1 = (-1)^n a_{n-1} I$$
$$\vdots$$
$$-B_{n-2} + AB_{n-1} = (-1)^n a_1 I$$
$$-B_{n-1} = (-1)^n I$$

Premultiplying these equations by $I, A, \ldots, A^n$ respectively and adding gives

$$0 = (-1)^n (a_n + a_{n-1}A + \ldots + A^n) = \chi_A(A)$$ ∎

The above results lead immediately to the following useful

COROLLARY *The minimum polynomial of a matrix divides its characteristic polynomial.*

163

PROOF   This follows from the above theorem and Lemma 6.15.   ∎

EXAMPLES

1. If $A = \text{diag}(1, 0, -1)$, then the characteristic polynomial of A is $\lambda(\lambda - 1)(\lambda + 1)$. By calculating $A^2 = \text{diag}(1, 0, 1)$ and $A^3 = \text{diag}(1, 0, -1)$ it is clear that no linear relation of lower degree than $A^3 - A = 0$ is true. Thus the minimum polynomial and characteristic polynomial coincide in this case.

2. If $A = \begin{pmatrix} 0 & 0 & 1 \\ 0 & 0 & 0 \\ 0 & 0 & 0 \end{pmatrix}$, then the characteristic polynomial of A is x^3 and since $A^2 = 0, A \neq 0$, the minimum polynomial is x^2. Thus, the characteristic polynomial and minimum polynomial need not be the same in general.

THEOREM 6.18   *The distinct linear factors of the minimum polynomial coincide with those of the characteristic polynomial.*

PROOF   Let the characteristic polynomial of A be

$$\chi_A(x) = (-1)^n (x - \lambda_1)^{m_1} (x - \lambda_2)^{m_2} \ldots (x - \lambda_k)^{m_k}$$

where $\lambda_1, \lambda_2, \ldots, \lambda_k$ are the distinct eigenvalues of A. Then by the corollary to Theorem 6.17,

$$m(x) = (x - \lambda_1)^{\ell_1} (x - \lambda_2)^{\ell_2} \ldots (x - \lambda_k)^{\ell_k}$$

where $0 \leqslant \ell_i \leqslant m_i$ $(i = 1, \ldots, k)$. We must show that $\ell_i > 0$ for all $i = 1, \ldots, k$.

Suppose that for some $1 \leqslant j \leqslant k$, $\ell_j = 0$, then $m(\lambda_j) \neq 0$. But, if λ is an eigenvalue of A, then there exists a non-zero $X \in V_n(K)$, such that

$$AX = \lambda X$$

and by induction

$$A^k X = \lambda^k X$$

for $k = 1, 2, \ldots$ and furthermore

$$m(A)X = m(\lambda)X$$

Thus, if $m(\lambda) \neq 0$ for some eigenvalue λ this would mean that $m(A) \neq 0$ which would contradict the definition of the minimum polynomial. Hence, no such eigenvalue λ_j can exist and $\ell_j > 0$ for all $j = 1, \ldots, k$.   ∎

COROLLARY *If a matrix $A \in M_n(K)$ has n distinct eigenvalues, then its minimum and characteristic polynomials coincide.*

We can now state an important criterion for a matrix to be diagonalizable in terms of its minimum polynomial. A proof is not included.

THEOREM 6.19 *A matrix $A \in M_n(K)$ with k distinct eigenvalues $\lambda_1, \ldots, \lambda_k$ is diagonalizable if and only if its minimum polynomial is $(x - \lambda_1) \ldots (x - \lambda_k)$.*

EXAMPLE

1. If $A = \begin{pmatrix} 1 & 1 & 0 & 0 \\ 0 & 1 & 0 & 0 \\ 0 & 0 & 2 & 0 \\ 0 & 0 & 0 & 2 \end{pmatrix}$ then its characteristic polynomial is clearly

$(x - 1)^2 (x - 2)^2$. By Theorem 6.18, the minimum polynomial of A must be one of (i) $(x - 1)(x - 2)$; (ii) $(x - 1)^2(x - 2)$ (iii) $(x - 1)(x - 2)^2$; (iv) $(x - 1)^2(x - 2)^2$.
The minimum polynomial is now determined by systematically eliminating these possibilities.

$$(A - I)(A - 2I) = \begin{pmatrix} 0 & 1 & 0 & 0 \\ 0 & 0 & 0 & 0 \\ 0 & 0 & 1 & 0 \\ 0 & 0 & 0 & 1 \end{pmatrix} \begin{pmatrix} -1 & 1 & 0 & 0 \\ 0 & -1 & 0 & 0 \\ 0 & 0 & 0 & 0 \\ 0 & 0 & 0 & 0 \end{pmatrix}$$

$$= \begin{pmatrix} 0 & -1 & 0 & 0 \\ 0 & 0 & 0 & 0 \\ 0 & 0 & 0 & 0 \\ 0 & 0 & 0 & 0 \end{pmatrix} \neq 0$$

$$(A - I)^2(A - 2I) = \begin{pmatrix} 0 & 1 & 0 & 0 \\ 0 & 0 & 0 & 0 \\ 0 & 0 & 1 & 0 \\ 0 & 0 & 0 & 1 \end{pmatrix} \begin{pmatrix} 0 & -1 & 0 & 0 \\ 0 & 0 & 0 & 0 \\ 0 & 0 & 0 & 0 \\ 0 & 0 & 0 & 0 \end{pmatrix} = 0$$

Thus the minimum polynomial is $(x - 1)^2(x - 2)$ and the matrix A is not diagonalizable.

2. If $A = \begin{pmatrix} 3 & 2 & 4 \\ 2 & 0 & 2 \\ 4 & 2 & 3 \end{pmatrix}$, then the characteristic polynomial is

$-(x+1)^2(x-8)$. We see that

$$(A+I)(A-8I) = \begin{pmatrix} 4 & 2 & 4 \\ 2 & 1 & 2 \\ 4 & 2 & 4 \end{pmatrix} \begin{pmatrix} -5 & 2 & 4 \\ 2 & -8 & 2 \\ 4 & 2 & -5 \end{pmatrix} = 0$$

and thus the minimum polynomial of A is $(x+1)(x-8)$ and so the matrix A is diagonalizable as we have seen in the example on page 155.

Exercises 6.4

1. Find the minimum polynomial of the following matrices

(i) $\begin{pmatrix} 1 & -1 \\ -1 & 1 \end{pmatrix}$ (ii) $\begin{pmatrix} 1 & 1 & 0 \\ 0 & 1 & 0 \\ 0 & 0 & 2 \end{pmatrix}$ (iii) $\begin{pmatrix} 1 & 0 & 0 \\ 0 & 1 & 1 \\ 0 & 0 & 2 \end{pmatrix}$

(iv) $\begin{pmatrix} 1 & -4 & 2 & 4 \\ 2 & -1 & -1 & 2 \\ 0 & -4 & 3 & 4 \\ 2 & 0 & -2 & 1 \end{pmatrix}$

2. Find the minimum polynomial and characteristic polynomial of the matrix

$$\begin{pmatrix} 0 & 1 & 0 & \ldots & 0 \\ 0 & 0 & 1 & \ldots & 0 \\ \cdot & \cdot & \cdot & & \cdot \\ \cdot & \cdot & \cdot & & \cdot \\ \cdot & \cdot & \cdot & & \cdot \\ 0 & 0 & 0 & & 1 \\ a_0 & a_1 & a_2 & \ldots & a_{n-1} \end{pmatrix}$$

Given a polynomial $f(x)$ of degree n, find an $n \times n$ matrix with $f(x)$ as its minimum polynomial.

3. If a matrix A has characteristic polynomial $(x-1)^3(x+2)^2(x-3)$, find all the possible minimum polynomials of A. Find matrices A which have these polynomials as their minimum polynomials.

166

4. Give a direct proof of the Cayley-Hamilton Theorem for
   (i) diagonalizable matrices, (ii) triangular matrices.

5. By calculating the minimum polynomial, determine which of the following matrices are diagonalizable:

(i) $\begin{pmatrix} 1 & -1 & -1 \\ 0 & 3 & 2 \\ 0 & -1 & 0 \end{pmatrix}$
(ii) $\begin{pmatrix} 2 & 0 & -1 \\ -1 & 2 & 2 \\ 1 & -1 & -1 \end{pmatrix}$

(iii) $\begin{pmatrix} 1 & 4 & -2 & -4 \\ 0 & 1 & 0 & -2 \\ 0 & 4 & -1 & -4 \\ 0 & 0 & 0 & -1 \end{pmatrix}$
(iv) $\begin{pmatrix} 3 & -4 & 0 & 5 \\ 1 & -1 & 0 & 1 \\ 2 & -3 & 1 & 4 \\ 0 & 0 & 0 & 1 \end{pmatrix}$

6.5 The Diagonalization of Symmetric Matrices

Let $M_n^{(s)}(\mathbf{R})$ denote the set of all real symmetric $n \times n$ matrices. We shall show that in this case, the problem considered in Section 6.3 can be solved completely.

DEFINITION 6.20   *$A, B \in M_n^{(s)}(\mathbf{R})$ are said to be **orthogonally similar** if there exists an orthogonal matrix $P \in M_n(\mathbf{R})$ such that*

$$B = P^t A P$$

(note that if P is orthogonal $P^t P = P P^t = I_n$ and so $P^{-1} = P^t$).

We first prove two lemmas.

LEMMA 6.21   *The eigenvalues of a real symmetric matrix are all real.*

PROOF   Let $A \in M_n^{(s)}(\mathbf{R})$ and λ be an eigenvalue of A with corresponding eigenvector $X \in V_n(\mathbf{R})$, then

$$AX = \lambda X$$

Taking complex conjugates and then transposes we obtain

$$\bar{A}\bar{X} = \bar{\lambda}\bar{X}$$

and

$$\bar{X}^t \bar{A}^t = \bar{\lambda}\bar{X}^t$$

Thus, we have

$$\bar{X}^t A X = \bar{\lambda} \bar{X}^t X = \lambda \bar{X}^t X$$

from which it follows that

$$(\lambda - \bar{\lambda})(\bar{X}^t X) = 0$$

Now, if $X = (x_1, x_2, \ldots, x_n)$, then
$\bar{X}^t X = \bar{x}_1 x_1 + \ldots + \bar{x}_n x_n = |x_1|^2 + \ldots + |x_n|^2 \neq 0$ and thus $\lambda = \bar{\lambda}$
and λ is real. ∎

LEMMA 6.22 *If the eigenvalues of a real symmetric matrix A are all distinct then the corresponding eigenvectors are orthogonal to each other.*

PROOF Let $\lambda_1, \lambda_2, \ldots, \lambda_n$ be the distinct eigenvalues of A with corresponding eigenvectors $X_1, X_2, \ldots, X_n$, i.e.
i.e. $AX_i = \lambda_i X_i$ $(i = 1, 2, \ldots, n)$. We wish to show that if $i \neq j$,
$(X_i, X_j) = X_i^t X_j = 0$.
We have that

$$X_i^t A = \lambda_i X_i^t$$

and

$$X_i^t A X_j = \lambda_i X_i^t X_j = \lambda_j X_i^t X_j$$

or

$$(\lambda_i - \lambda_j) X_i^t X_j = 0$$

from which it follows that $X_i^t X_j = 0$ since $\lambda_i - \lambda_j \neq 0$. ∎
From these two results it now follows that

THEOREM 6.23 *If A is a real symmetric matrix with distinct eigenvalues then A is orthogonally similar to a diagonal matrix.*

PROOF Let $\lambda_1, \lambda_2, \ldots, \lambda_n$ be the eigenvalues of A and $X_1, X_2, \ldots, X_n$
the corresponding eigenvectors. Let $Y_i = \dfrac{1}{|X_i|} X_i$ $(i = 1, 2, \ldots, n)$ and

put $P = (Y_1, Y_2, \ldots, Y_n)$, then P is an orthogonal matrix, since by
Lemma 6.22 above $(Y_i, Y_j) = 0$ $(i \neq j)$ and $(Y_i, Y_i) = 1$ $(i, j = 1, 2, \ldots, n)$.
Furthermore, $Y_1, Y_2, \ldots, Y_n$ are also eigenvectors corresponding to the
eigenvalues $\lambda_1, \lambda_2, \ldots, \lambda_n$ and by Theorem 6.11

$$P^t A P = P^{-1} A P = \operatorname{diag}(\lambda_1, \lambda_2, \ldots, \lambda_n)$$ ∎

In the next theorem, we see that this result is even true when the eigenvalues are not distinct.

THEOREM 6.24   *Let A be a real symmetric $n \times n$ matrix. Then, there exists an orthogonal matrix P such that $P^t A P = diag\,(\lambda_1, \lambda_2, \ldots, \lambda_n)$ where $\lambda_1, \lambda_2, \ldots, \lambda_n$ are the eigenvalues of A (possibly repeated).*

PROOF   The proof is by induction on n.
If $n = 1$, the theorem is trivially true.
If $n > 1$, we assume that every $(n-1) \times (n-1)$ real symmetric matrix is orthogonally similar to a diagonal matrix.

Let λ be a eigenvalue of A and X the corresponding eigenvector, thus $AX = \lambda X$. Put $X_1 = \dfrac{1}{|X|} X$, then by the Gram-Schmidt orthogonalization procedure (see Theorem 5.6) an orthonormal **R**-basis $\{X_1, X_2, \ldots, X_n\}$ for $V_n(\mathbf{R})$ may be constructed.
Put $U = (X_1, X_2, \ldots, X_n)$, then U is an orthogonal $n \times n$ matrix and

$$U^t A U = \begin{pmatrix} \lambda(X_1, X_1) & * \\ \lambda(X_2, X_1) & \\ \cdot & B \\ \cdot & \\ \lambda(X_n, X_1) & \end{pmatrix}$$

$$= \begin{pmatrix} \lambda & * \\ 0 & \\ \cdot & B \\ \cdot & \\ 0 & \end{pmatrix}$$

where B is an $(n-1) \times (n-1)$ matrix. Since A is a symmetric matrix and

$$(U^t A U)^t = U^t A U$$

then $U^t A U$ is a symmetric matrix and B is also a symmetric matrix and

$$U^t A U = \begin{pmatrix} \lambda & 0 \\ \hline 0 & B \end{pmatrix}$$

Now, by the induction assumption, there exists an $(n-1) \times (n-1)$ orthogonal matrix V such that

$$V^t B V = \operatorname{diag}(\lambda_2, \ldots, \lambda_n)$$

where $\lambda_2, \ldots, \lambda_n$ are eigenvalues of B, and hence also eigenvalues of A. Let

$$V' = \left(\begin{array}{c|c} 1 & 0 \\ \hline 0 & V \end{array} \right)$$

then V' is an $n \times n$ orthogonal matrix and

$$V'^t U^t A U V' = V'^t \left(\begin{array}{c|c} \lambda & 0 \\ \hline 0 & B \end{array} \right) V'$$

$$= \operatorname{diag}(\lambda, \lambda_2, \ldots, \lambda_n)$$

and UV' is an orthogonal matrix, which completes the proof. ∎

EXAMPLES

1. If $A = \begin{pmatrix} 10 & -14 & -10 \\ -14 & 7 & -4 \\ -10 & -4 & 19 \end{pmatrix}$, then the characteristic polynomial of A is

$$\chi_A(x) = \begin{vmatrix} 10-x & -14 & -10 \\ -14 & 7-x & -4 \\ -10 & -4 & 19-x \end{vmatrix}$$

which on evaluation gives

$$\chi_A(x) = (x-18)(x-27)(x+9)$$

and the eigenvalues are $\lambda = 18, 27$ and -9.

If $\lambda = 18$ then $\begin{pmatrix} -8 & -14 & -10 \\ -14 & -11 & -4 \\ -10 & -4 & 1 \end{pmatrix} \begin{pmatrix} 1 \\ -2 \\ 2 \end{pmatrix} = \begin{pmatrix} 0 \\ 0 \\ 0 \end{pmatrix}$

if $\lambda = -9$, then $\begin{pmatrix} 19 & -14 & -10 \\ -14 & 16 & -4 \\ -10 & -4 & 28 \end{pmatrix} \begin{pmatrix} 2 \\ 2 \\ 1 \end{pmatrix} = \begin{pmatrix} 0 \\ 0 \\ 0 \end{pmatrix}$

and if $\lambda = 27$, then $\begin{pmatrix} -17 & -14 & -10 \\ -14 & -20 & -4 \\ -10 & -4 & -8 \end{pmatrix} \begin{pmatrix} -2 \\ 1 \\ 2 \end{pmatrix} = \begin{pmatrix} 0 \\ 0 \\ 0 \end{pmatrix}$

170

If we now let $P = \frac{1}{3}\begin{pmatrix} 1 & 2 & -2 \\ -2 & 2 & 1 \\ 2 & 1 & 2 \end{pmatrix}$ then P is an orthogonal matrix and

$$P^t A P = \text{diag}(18, -9, 27)$$

2. If $A = \begin{pmatrix} 7 & -1 & -2 \\ -1 & 7 & 2 \\ -2 & 2 & 10 \end{pmatrix}$ then the characteristic polynomial of A is

$$\chi_A(x) = \begin{vmatrix} 7-x & -1 & -2 \\ -1 & 7-x & 2 \\ -2 & 2 & 10-x \end{vmatrix} = (x-6)^2(x-12);$$

the eigenvalues of A are $\lambda = 6$ and $\lambda = 12$. When $\lambda = 12$, then

$$\begin{pmatrix} -5 & -1 & -2 \\ -1 & -5 & 2 \\ -2 & 2 & -2 \end{pmatrix} \begin{pmatrix} -1 \\ 1 \\ 2 \end{pmatrix} = 0$$

and when $\lambda = 6$, then

$$\begin{pmatrix} 1 & -1 & -2 \\ -1 & 1 & 2 \\ -2 & 2 & 4 \end{pmatrix} \begin{pmatrix} 1 & 0 \\ 1 & 2 \\ 0 & -1 \end{pmatrix} = 0$$

The two vectors $(1, 1, 0)$ and $(0, 2, -1)$ are not orthogonal to each other, but by applying the Gram-Schmidt orthogonalization procedure we obtain

$$y_1 = (1, 1, 0)$$
$$y_2 = (0, 2, -1) - (1, 1, 0) = (-1, 1, -1)$$

Then $(1, 1, 0)$ and $(1, -1, 1)$ are orthogonal vectors which are eigenvectors corresponding to the eigenvalue $\lambda = 6$. Now, by normalizing each vector and putting

$$P = \begin{pmatrix} \dfrac{-1}{\sqrt{6}} & \dfrac{1}{\sqrt{2}} & \dfrac{1}{\sqrt{3}} \\[2mm] \dfrac{1}{\sqrt{6}} & \dfrac{1}{\sqrt{2}} & \dfrac{-1}{\sqrt{3}} \\[2mm] \dfrac{2}{\sqrt{6}} & 0 & \dfrac{1}{\sqrt{3}} \end{pmatrix} = \dfrac{1}{\sqrt{6}} \begin{pmatrix} -1 & \sqrt{3} & \sqrt{2} \\ 1 & \sqrt{3} & -\sqrt{2} \\ 2 & 0 & \sqrt{2} \end{pmatrix}$$

we have that $P^t AP = \text{diag}\,(12, 6, 6)$.

Exercises 6.5

1. Find an orthogonal matrix U such that $U^t AU$ is a diagonal matrix for the following matrices A

(i) $\begin{pmatrix} 1 & 0 & -4 \\ 0 & 5 & 4 \\ -4 & 4 & 3 \end{pmatrix}$ (ii) $\begin{pmatrix} 11 & 0 & 6 \\ 0 & 5 & 6 \\ 6 & 6 & -2 \end{pmatrix}$ (iii) $\begin{pmatrix} 7 & -1 & -2 \\ -1 & 7 & 2 \\ -2 & 2 & 10 \end{pmatrix}$

(iv) $\begin{pmatrix} 18 & -1 & -4 \\ -1 & 18 & -4 \\ -4 & -4 & 3 \end{pmatrix}$

6.6 Quadratic Forms

Throughout this section we assume that V is a finite dimensional **R**-space.

DEFINITION 6.25 *If $\mathscr{B} = \{v_1, v_2, \ldots, v_n\}$ is a R-basis for V then a* **real quadratic form** *on V is a function $Q : V \to \mathbf{R}$ defined by*

$$Q(v) = \sum_{i=1}^{n} \sum_{j=1}^{n} a_{ij}\, x_i\, x_j \tag{1}$$

where $v = \sum\limits_{i=1}^{n} x_i\, v_i\ (x_i \in R)$ and $A = (a_{ij})$ is a real symmetric matrix.

For example, if $A = \begin{pmatrix} 2 & 1 \\ 1 & 3 \end{pmatrix}$, then $Q : V \to \mathbf{R}$ is given by

172

$$Q(v) = 2x_1{}^2 + 3x_2{}^2 + x_1 x_2 + x_2 x_1$$

$$= 2x_1{}^2 + 3x_2{}^2 + 2x_1 x_2$$

The expression (1) will usually be written as

$$Q(v) = \sum_{\substack{i,j=1 \\ i \leqslant j}}^{n} a_{ij} x_i x_j$$

and the coefficient a_{ij} in this expression is shared equally between the (i,j)- and (j,i)-positions of the matrix A.

If $X = (x_1, x_2, \ldots, x_n)$ is the co-ordinate vector of v, then $Q(v)$ can be expressed in matrix form as

$$Q(v) = XAX^t$$

We now consider the effect of a change of **R**-basis on this expression for Q.

Let $\mathscr{B}' = \{v_1', v_2', \ldots, v_n'\}$ be another **R**-basis for V, then (see p.100)
$v_j' = \sum_{i=1}^{n} p_{ij} v_i$ $(j = 1, \ldots, n)$, where $P = (p_{ij})$ is an invertible matrix.

If $v = \sum_{j=1}^{n} y_j v_j'$, then

$$v = \sum_{j=1}^{n} y_j \left(\sum_{i=1}^{n} p_{ij} v_i \right)$$

$$= \sum_{i=1}^{n} \left(\sum_{j=1}^{n} p_{ij} y_j \right) v_i$$

and hence

$$X^t = PY^t \quad \text{where} \quad Y = (y_1, \ldots, y_n)$$

Hence,

$$Q(v) = XAX^t = YP^t APY^t = Y(P^t AP) Y^t$$

where $P^t AP$ is also a symmetric matrix since $(P^t AP)^t = P^t AP$.

If $D = \text{diag}(\lambda_1, \lambda_2, \ldots, \lambda_n)$ then $Q : V \to \mathbf{R}$ defined by

$$Q(v) = \lambda_1 x_1{}^2 + \ldots + \lambda_n x_n{}^2$$

is a real quadratic form on V which has a simple form called a **diagonal quadratic form**. We consider whether it is possible to find a **R**-basis for V so that a given real quadratic form can be represented as a diagonal quadratic form. We shall be even more restrictive with a view to later geometrical applications; we shall insist that the matrix P involved in the transformation is orthogonal. We now of course have precisely the same problem as was considered in the previous section and it follows immediately from Theorem 6.24 that we have

THEOREM 6.26 *Any real quadratic form may be reduced to a diagonal quadratic form by means of an orthogonal transformation.*

This result has a useful and elegant application in analytic geometry. The general equation of a conic section is of the form

$$ax^2 + 2bxy + cy^2 + ux + vy + p = 0 \tag{1}$$

We call

$$ax^2 + 2bxy + cy^2$$

the quadratic form associated with (1). Equation (1) can therefore be expressed in the matrix form

$$(x,y) \begin{pmatrix} a & b \\ b & c \end{pmatrix} \begin{pmatrix} x \\ y \end{pmatrix} + (u,v) \begin{pmatrix} x \\ y \end{pmatrix} + p = 0$$

Thus, if $X = (x,y)$, $A = \begin{pmatrix} a & b \\ b & c \end{pmatrix}$, $T = (u,v)$, we have

$$XAX^t + TX^t + p = 0$$

The matrix A is symmetric and by the above theorem, there exists an orthogonal matrix P such that

$$P^tAP = \begin{pmatrix} \lambda_1 & 0 \\ 0 & \lambda_2 \end{pmatrix}$$

where λ_1, λ_2 are the eigenvalues of A. Thus, by carrying out the transformation $\begin{pmatrix} x \\ y \end{pmatrix} = P\begin{pmatrix} x' \\ y' \end{pmatrix}$ (or $\begin{pmatrix} x' \\ y' \end{pmatrix} = P^t\begin{pmatrix} x \\ y \end{pmatrix}$) the form (1) reduces to

$$X'P^t APX'^t + TPX'^t + p = 0$$

or

$$\lambda_1 x'^2 + \lambda_2 y'^2 + u'x' + v'y' + p' = 0 \tag{2}$$

where $p' = p$ and $(u',v') = TP$.

Thus, by introducing new axes, called the *principal* axes $0x'$ and $0y'$ in the xy-plane in directions X_1 and X_2, where $P = (X_1,X_2)$ the conic section has a graph represented by (2). Equation 2 may be reduced further, depending on the values of λ_1 and λ_2. The following cases are considered.

174

Case 1 $\lambda_1 \neq 0, \lambda_2 \neq 0$

If we put
$$x'' = x' + \frac{v'}{2\lambda_1} \ , \ y'' = y' + \frac{u'}{2\lambda_2}$$

in (2), we obtain

$$\lambda_1 y''^2 + \lambda_2 y''^2 = p''$$

where

$$p'' = -\left(p' - \frac{u'v'}{4\lambda_1^2}\right)$$

If $\lambda_1 = \lambda_2$, then this is the equation of a *circle* with centre $\left(-\dfrac{u'}{2\lambda_1} \ , \ -\dfrac{v'}{2\lambda_1}\right)$ in the $x'y'$-plane and radius $\sqrt{p''/\lambda_1}$. If $\lambda_1 > 0, \lambda_2 > 0$ and $\lambda_1 \neq \lambda_2$, it is the equation of an *ellipse* with centre $\left(-\dfrac{u'}{2\lambda_1} \ , \ -\dfrac{v'}{2\lambda_2}\right)$ in the $x'y'$-plane which meets the new $0x''$-axis at the points $\left(\pm\sqrt{p''/\lambda_1} \ , 0\right)$ and the new $0y''$-axis at the points $\left(0, \pm\sqrt{p''/\lambda_2}\right)$.

If $\lambda_1 > 0, \lambda_2 < 0$, it is the equation of a *hyperbola* with centre $\left(-\dfrac{u'}{2\lambda_1} \ , \ -\dfrac{v'}{2\lambda_2}\right)$ in the $x'y'$-plane and the vertices are $\left(\pm\sqrt{p''/\lambda_1} \ , 0\right)$ and the two lines $y''/\sqrt{p''/\lambda_2} = \pm x''/\sqrt{p''/\lambda_1}$ (or $\sqrt{\lambda_2}y'' = \pm\sqrt{\lambda_1}x''$)

are the asymptotes.

Case 2 $\lambda_1 = 0, \ \lambda_2 \neq 0$

If we put
$$y'' = y' + \frac{v'}{2\lambda_2}$$

in (2), we obtain

$$\lambda_2 y''^2 + u'x' + p'' = 0$$

Further, putting $x'' = x' + p'/u'$, if $u' \neq 0$, this becomes

$$\lambda_2 y''^2 + u'x'' = 0$$

175

which is the equation of a *parabola* with vertex $\left(-\dfrac{p''}{u'}, -\dfrac{v'}{2\lambda_2}\right)$ in the

$x'y'$-plane and focus $(-u'/4\lambda_2, 0)$. The case $\lambda_1 \neq 0, \lambda_2 = 0$ can be dealt with similarly. There are further "degenerate" cases which we shall not consider further here.

The standard form of these conics are given in Appendix 1.

This is illustrated in the following examples. The first examples are "central" conics.

EXAMPLES

1. Draw a graph of the conic section

$$5x^2 - 6xy + 5y^2 = 8$$

In the matrix form this is

$$(x, y) \begin{pmatrix} 5 & -3 \\ -3 & 5 \end{pmatrix} \begin{pmatrix} x \\ y \end{pmatrix} = 8$$

The eigenvalues are roots of

$$\begin{vmatrix} 5-x & -3 \\ -3 & 5-x \end{vmatrix} = (2-x) \begin{vmatrix} 1 & -3 \\ 0 & 8-x \end{vmatrix} = (2-x)(8-x)$$

i.e. the eigenvalues are $\lambda = 2, 8$ and the corresponding eigenvectors are $\begin{pmatrix} 1 \\ 1 \end{pmatrix}$ and $\begin{pmatrix} 1 \\ -1 \end{pmatrix}$. Normalizing these vectors, we have

$$P = \frac{1}{\sqrt{2}} \begin{pmatrix} 1 & 1 \\ 1 & -1 \end{pmatrix}$$

and then

$$P^t A P = \begin{pmatrix} 2 & 0 \\ 0 & 8 \end{pmatrix}$$

Now taking new axes $0x', 0y'$ in the directions $\left(\dfrac{1}{\sqrt{2}}, \dfrac{1}{\sqrt{2}}\right)$ and

$\left(\dfrac{1}{\sqrt{2}}, \dfrac{-1}{\sqrt{2}}\right)$ respectively, $\left(\text{or putting } x = \dfrac{1}{\sqrt{2}}(x'+y'), y = \dfrac{1}{\sqrt{2}}(x'-y')\right)$
the equation of the conic becomes

$$2x'^2 + 8y'^2 = 8$$

or $\quad \dfrac{x'^2}{4} + y'^2 = 1$

That is, we have an ellipse with graph as in Fig. 2.

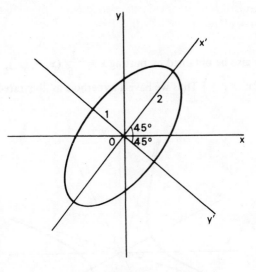

Figure 2

2. Draw a graph of the conic section

$$7x^2 - 12xy - 2y^2 = 3$$

The corresponding matrix is

$$A = \begin{pmatrix} 7 & -6 \\ -6 & -2 \end{pmatrix}$$

which has eigenvalues -5 and 10 and eigenvectors $\begin{pmatrix} 1 \\ 2 \end{pmatrix}$ and $\begin{pmatrix} -2 \\ 1 \end{pmatrix}$ respectively. Let $P = \dfrac{1}{\sqrt{5}} \begin{pmatrix} 1 & -2 \\ 2 & 1 \end{pmatrix}$, then

$$P^t A P = \begin{pmatrix} -5 & 0 \\ 0 & 10 \end{pmatrix}$$

177

Taking new axes Ox', Oy' in the directions $\left(\dfrac{1}{\sqrt5}, \dfrac{2}{\sqrt5}\right)$ and $\left(\dfrac{-2}{\sqrt5}, \dfrac{1}{\sqrt5}\right)$ respectively, the equation of the conic becomes

$$-5x'^2 + 10y'^2 = 10$$

or

$$-\frac{x'^2}{2} + y'^2 = 1$$

$\left(\text{This may also be obtained by putting } x = \dfrac{1}{\sqrt5}(x' - 2y'),\right.$
$y = \dfrac{1}{\sqrt5}(2x' + y').\Big)$ Thus we have a hyperbola as illustrated in Fig. 3.

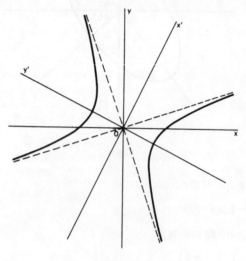

Figure 3

3. Identify and sketch the graph of the conic section whose equation is

$$x^2 + y^2 - 2xy - 4\sqrt2 x + 6 = 0$$

The matrix form is

$$(x, y) \begin{pmatrix} 1 & -1 \\ -1 & 1 \end{pmatrix} \begin{pmatrix} x \\ y \end{pmatrix} + (-4\sqrt2, 0) \begin{pmatrix} x \\ y \end{pmatrix} + 6 = 0$$

178

The characteristic polynomial of this matrix is $x(x-2)$ and the eigenvalues are 0 and 2. The corresponding eigenvectors are $\begin{pmatrix} 1 \\ 1 \end{pmatrix}$ and $\begin{pmatrix} 1 \\ -1 \end{pmatrix}$ respectively. If we now let

$$P = \frac{1}{\sqrt{2}} \begin{pmatrix} 1 & 1 \\ 1 & -1 \end{pmatrix}$$

then, taking new axes $0x'$, $0y'$ in the directions $(1/\sqrt{2}, 1/\sqrt{2})$ and $(1/\sqrt{2}, -1/\sqrt{2})$ respectively the equation of the conic section becomes

$$2y'^2 + 1/\sqrt{2}(-4\sqrt{2}, 0) \begin{pmatrix} 1 & 1 \\ 1 & -1 \end{pmatrix} \begin{pmatrix} x' \\ y' \end{pmatrix} + 6 = 0$$

or

$$y'^2 - 2y' - 2x' + 3 = 0$$

from which we obtain

$$(y' - 1)^2 = 2(x' - 1)$$

Thus, the conic section is a parabola with vertex $(1,1)$ in (x', y')-plane, which has coordinates $1/\sqrt{2} \begin{pmatrix} 1 & 1 \\ 1 & -1 \end{pmatrix} \begin{pmatrix} 1 \\ 1 \end{pmatrix} = \begin{pmatrix} \sqrt{2} \\ 0 \end{pmatrix}$, i.e. $(\sqrt{2}, 0)$ in the (x, y)-plane as illustrated in Figure 4.

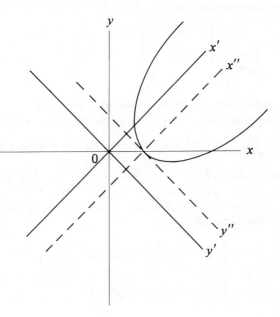

The same method can be used for quadric surfaces in 3-space. The general equation of a quadric surface is of the form

$$ax^2 + by^2 + cz^2 + 2dxy + 2exz + 2fyz + ux + vy + wz + p = 0$$

which may be again represented in the matrix form

$$XAX^t + TX^t + p = 0$$

where now

$$A = \begin{pmatrix} a & d & e \\ d & b & f \\ e & f & c \end{pmatrix}, \quad X = (x,y,z), \quad T = (u,v,w)$$

This may be reduced in precisely the same way as the conic section — details are not given in this case. In Appendix 2, the graphs of the non-degenerate surfaces are given. The methods are illustrated in the following examples.

EXAMPLES

1. Identify the quadric surface whose equation is

$$7x^2 + 7y^2 + 10z^2 - 2xy - 4xz + 4yz = 24$$

The corresponding matrix is

$$A = \begin{pmatrix} 7 & -1 & -2 \\ -1 & 7 & 2 \\ -2 & 2 & 10 \end{pmatrix}$$

which has been considered in Example 2 on p.168.
The eigenvalues are $\lambda = 6, 12$. Corresponding to $\lambda = 12$ the eigenvector is $\left(-\dfrac{1}{\sqrt{6}}, \dfrac{1}{\sqrt{6}}, \dfrac{2}{\sqrt{6}}\right)$ and corresponding to $\lambda = 6$ the orthogonal eigenvectors are $\left(\dfrac{1}{\sqrt{2}}, \dfrac{1}{\sqrt{2}}, 0\right)$ and $\left(\dfrac{1}{\sqrt{3}}, \dfrac{-1}{\sqrt{3}}, \dfrac{1}{\sqrt{3}}\right)$. Taking new axes $0x', 0y'$ and $0z'$ in the direction of these unit vectors respectively (or equivalently by substituting

$$x = \frac{1}{\sqrt{6}} (-x' + \sqrt{3}\, y' + \sqrt{2}\, z')$$

$$y = \frac{1}{\sqrt{6}} (x' + \sqrt{3}\, y' - \sqrt{2}\, z')$$

$$z = \frac{1}{\sqrt{6}} (2x' \qquad\quad + \sqrt{2}\, z'))$$

180

the quadric surface takes the form,

$$12x'^2 + 6y'^2 + 6z'^2 = 24$$

or

$$\frac{x'^2}{2} + \frac{y'^2}{4} + \frac{z'^2}{4} = 1$$

which is the equation of an ellipsoid.

2. Identify the quadric surface whose equation is

$$2x^2 + y^2 - 3z^2 + 12xy - 4xz - 8yz + 8x + 12y + 22z + 44 = 0$$

The corresponding matrix form is

$$(x,y,z) \begin{pmatrix} 2 & 6 & -2 \\ 6 & 1 & -4 \\ -2 & -4 & -3 \end{pmatrix} \begin{pmatrix} x \\ y \\ z \end{pmatrix} + (8,12,22) \begin{pmatrix} x \\ y \\ z \end{pmatrix} + 44 = 0$$

It may be verified that the eigenvalues of this matrix are -3, -6 and 9 and the corresponding eigenvectors are

$$\begin{pmatrix} 2 \\ -1 \\ 2 \end{pmatrix}, \begin{pmatrix} -1 \\ 2 \\ 2 \end{pmatrix} \text{ and } \begin{pmatrix} 2 \\ 2 \\ -1 \end{pmatrix}$$

respectively. If we put

$$P = \tfrac{1}{3} \begin{pmatrix} 2 & -1 & 2 \\ -1 & 2 & 2 \\ 2 & 2 & -1 \end{pmatrix}$$

and take new axes $0x'$, $0y'$, $0z'$ in the direction of the unit vectors $\frac{1}{3}(2,-1,2)$, $\frac{1}{3}(-1,2,2)$, $\frac{1}{3}(2,2,-1)$ respectively, then the equation of the quadric surface becomes

$$3x'^2 + 6y'^2 - 9z'^2 - 16x' - 20y' - 6z' - 44 = 0$$

$$\left(\text{since } \tfrac{1}{3}(8,12,22) \begin{pmatrix} 2 & -1 & 2 \\ -1 & 2 & 2 \\ 2 & 2 & -1 \end{pmatrix} = (16,20,6) \right)$$

Now, put $x'' = x' - 8/3, y'' = y' - 5/3, z'' = z' + 1/3$; then the equation of the quadric surface becomes

$$x''^2 + 2y''^2 - 3z''^2 = 27$$

or

$$\frac{x''^2}{(3\sqrt{3})^2} + \frac{y''^2}{\left(\frac{3\sqrt{3}}{\sqrt{2}}\right)^2} - \frac{z''^2}{3^2} = 1$$

181

Thus, we have a hyperboloid of one sheet (see Appendix 2) with centre

$$1/9(8,5,-1) \begin{pmatrix} 2 & -1 & 2 \\ -1 & 2 & 2 \\ 2 & 2 & -1 \end{pmatrix} = (1,0,3)$$

with axes in the direction of the unit vectors given above.

Exercise 6.6

1. Reduce the following, real quadratic forms to diagonal form
   (i) $x^2 + y^2 + xy$
   (ii) $x^2 + y^2 - xy$
   (iii) $3x^2 + 3y^2 + 5z^2 - 2xy$
   (iv) $2xy + 2xz + 2yz$
   (v) $5x^2 + 11y^2 - 2z^2 + 12xz + 12yz$

2. Find the principal axes, centre and sketch the graph of the following conics
   (i) $xy = 2$
   (ii) $3x^2 - 2y^2 + 12xy = 42$
   (iii) $7x^2 + 4y^2 - 4xy = 24$
   (iv) $13x^2 + 13y^2 + 10xy = 72.$
   (v) $5x^2 + 26xy + 5y^2 - 70x - 38y = 7$
   (vi) $73x^2 + 72xy + 52y^2 - 190x - 80y = -25$

3. Find the principal axes, centre and identify the quadric surface whose equation is
   (i) $2xy + 2xz + 2yz = 4$
   (ii) $x^2 + 2y^2 + z^2 - 2xy - 2yz = 1$
   (iii) $4x^2 + 3y^2 + 3z^2 - 4xy + 4xz - 6yz = 16$
   (iv) $5x^2 + 2y^2 + 2z^2 + 4xy + 4xz + 2yz = 7$
   (v) $x^2 + y^2 + z^2 - 4xy - 4xz - 4yz = 3$
   (vi) $11x^2 + 18y^2 + 4z^2 - 12xy + 12xz = 22.$
   (vii) $7x^2 - 2y^2 + 4z^2 + 16yz - 20xz - 4xy + 5x + 5y - 2z = 0$
   (viii) $7x^2 - 2y^2 - 2z^2 - 8yz + 10xz - 10xy + 2x + 2z = 0$

APPENDIX 1

Conic Sections

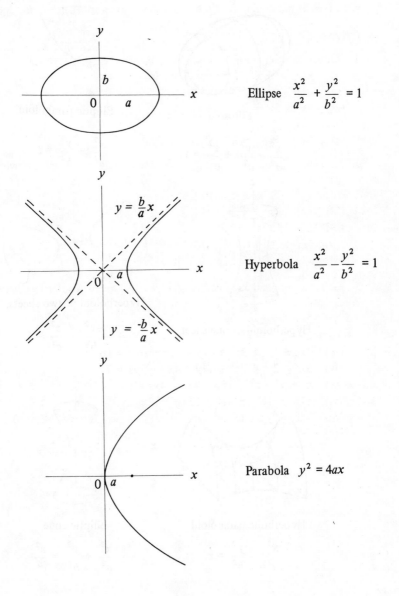

Ellipse $\dfrac{x^2}{a^2} + \dfrac{y^2}{b^2} = 1$

Hyperbola $\dfrac{x^2}{a^2} - \dfrac{y^2}{b^2} = 1$

Parabola $y^2 = 4ax$

Quadric Surfaces

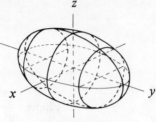

Ellipsoid

$$\frac{x^2}{a^2} + \frac{y^2}{b^2} + \frac{z^2}{c^2} = 1$$

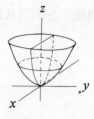

Elliptic paraboloid

$$\frac{x^2}{a^2} + \frac{y^2}{b^2} = z$$

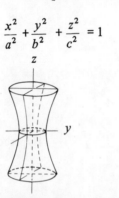

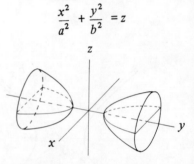

Hyperboloid of two sheets

$$\frac{x^2}{a^2} - \frac{y^2}{b^2} - \frac{z^2}{c^2} = 1$$

Hyperboloid of one sheet

$$\frac{x^2}{a^2} + \frac{y^2}{b^2} - \frac{z^2}{c^2} = 1$$

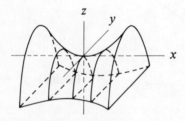

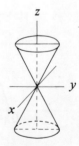

Hyperbolic paraboloid

$$\frac{x^2}{a^2} - \frac{y^2}{b^2} = z$$

Elliptic cone

$$\frac{x^2}{a^2} + \frac{y^2}{b^2} = \frac{z^2}{c^2}$$

Solutions to Exercises

Exercises 1.2

1. (i) $\begin{pmatrix} 1 & 0 & 0 & 5/8 \\ 0 & 1 & 0 & -1/8 \\ 0 & 0 & 1 & 1/8 \end{pmatrix}$ 　　(ii) $\begin{pmatrix} 1 & 0 & 0 \\ 0 & 1 & 0 \\ 0 & 0 & 1 \\ 0 & 0 & 0 \end{pmatrix}$

(iii) $\begin{pmatrix} 1 & 0 & 1/2 & 1/2 \\ 0 & 1 & 2 & 1 \\ 0 & 0 & 0 & 0 \\ 0 & 0 & 0 & 0 \end{pmatrix}$ 　(iv) $\begin{pmatrix} 1 & 0 & 1 & 0 & 2 & 0 \\ 0 & 1 & 1 & 0 & 0 & 1 \\ 0 & 0 & 0 & 1 & 2 & 1 \\ 0 & 0 & 0 & 0 & 0 & 0 \\ 0 & 0 & 0 & 0 & 0 & 0 \end{pmatrix}$

(v) $\begin{pmatrix} 1 & 0 & 1+i & 1 \\ 0 & 1 & (1+i)/2 & (1-i)/2 \\ 0 & 0 & 0 & 0 \end{pmatrix}$ (vi) $\begin{pmatrix} 1 & 0 & 0 & 0 \\ 0 & 1 & 0 & 0 \\ 0 & 0 & 1 & 0 \\ 0 & 0 & 0 & 1 \end{pmatrix}$

2. (i)　Yes 　　　　　(ii)　No

Exercise 1.3

1. General solutions are:

(i) $(-\lambda-2\mu-3\nu, 3\lambda-4\mu-\nu, 5\lambda, 5\mu, 5\nu)$

(ii) $(-9\lambda-7\mu, 19\lambda+35\mu, 7\lambda-49\mu, 14\lambda, 14\mu)$

(iii) $(\lambda, -\lambda, 4\lambda, 3\lambda)$

(iv) $(3\lambda-6\mu, 7\lambda-9\mu+2\nu, 3\lambda, 3\mu, 3\nu, 3\nu)$

where λ, μ, ν are arbitrary.

185

2. General solutions are:

   (i) $(4-2\lambda-2\mu, 9-4\lambda-3\mu, \lambda, \mu)$

   (ii) No solution

   (iii) $(-1-\lambda+\mu, 7/2+1/2\lambda-5/2\mu, -1/2-1/2\lambda+1/2\mu, \lambda, \mu)$

   (iv) No solution

   where λ, μ are arbitrary.

3. (a) (i) $\lambda \neq 8$; (ii) $\lambda = 8, \mu \neq 8$; (iii) $\lambda = 8, \mu = 8$

   (b) (i) $\lambda \neq 11$; (ii) $\lambda = 11, \mu \neq 3$; (iii) $\lambda = 11, \mu = 3$

4. $\lambda \neq 2, \lambda \neq 1$

   when $\lambda = 1$, $2a - 3b + c = 0$, $(2a-b-\mu, b-a, \mu)$

   when $\lambda = 2$, $2a-3b+c = 0$, $(b-a-\mu, \mu, -b+2a)$, μ arbitrary

5. (i) $\lambda = -14, (-1+\mu, 3-2\mu, \mu); \lambda \neq -14, (0, 1, 1)$

   (ii) $\lambda \neq 1/2; \lambda = -1, (-2+\mu, 1, \mu); \lambda \neq 1/2, \lambda \neq -1, \dfrac{1}{2(1-2\lambda)}$
   $(-14\lambda^2+5\lambda+12, 8\lambda^2-4\lambda-6, 3-2\lambda)$

   (iii) $\lambda = 1, (1/3, -2/3)$

   (iv) $\lambda^2+2\lambda = a, (-5a+2+\mu, 3a-1-\mu, \mu, a-1)$, μ arbitrary

6. $(-\lambda-2\mu, 1-\lambda, \lambda, 1-2\mu, \mu)$, λ, μ arbitrary

7. $n = 8, \lambda(1,-1,0,1,-1,0,1,-1)$, λ arbitrary

   $n = 9, (0,0,0,0,0,0,0,0,0)$

Exercises 1.4

3. $a = 2, b = 1/4, c = -1/64$ or $a = -2, b = -1/4, c = 1/64$

5. If $A = (\alpha_{ij})$ then $\alpha_{i,n-j+1} = \alpha_{n-i+1,j}$ $(i, j=1, \ldots, n)$

Exercises 1.5

4. $A(\alpha)^{-1} = A(-\alpha); x^3 - 3x^2 + 3x - 1 = 0$

6. $\pm \dfrac{1}{\sqrt{2}} \begin{pmatrix} 1 & 1 \\ 1 & -1 \end{pmatrix}$, $\pm \dfrac{1}{\sqrt{2}} \begin{pmatrix} 1 & 1 \\ -1 & 1 \end{pmatrix}$

Exercises 1.6

1. (i)
$$\frac{1}{2}\begin{pmatrix} -3 & 2 & 1 \\ -1 & 0 & 1 \\ 5 & -2 & -1 \end{pmatrix}$$

(ii)
$$\frac{1}{3}\begin{pmatrix} 11 & -9 & 1 \\ -7 & 9 & -2 \\ 2 & -3 & 1 \end{pmatrix}$$

(iii)
$$\frac{1}{75}\begin{pmatrix} 5 & -2 & 19 \\ 25 & 5 & -10 \\ 5 & 13 & -11 \end{pmatrix}$$

(iv)
$$\frac{1}{25}\begin{pmatrix} 2 & 3 & 1 \\ 17 & 13 & -29 \\ 9 & 1 & -8 \end{pmatrix}$$

(v)
$$\frac{1}{5}\begin{pmatrix} 3 & 2 & 10 & -3 \\ 0 & 0 & 5 & 5 \\ 1 & -1 & 0 & 4 \\ 3 & 2 & 5 & -3 \end{pmatrix}$$

(vi)   None

(vii)
$$\frac{1}{2}\begin{pmatrix} 2i & -2 & 1+i \\ -2-2i & -2i & -1+i \\ -1-i & 1-i & 0 \end{pmatrix}$$

Exercises 1.7

1. (i)
$$P = \frac{1}{3}\begin{pmatrix} 3 & 0 \\ -2 & 1 \end{pmatrix} \quad Q = \begin{pmatrix} 1 & 0 & 1 \\ 0 & 1 & -1 \\ 0 & 0 & 1 \end{pmatrix}$$

(ii)
$$P = \frac{1}{7}\begin{pmatrix} 3 & 2 & 1 \\ -1 & 4 & 2 \\ 2 & -1 & 3 \end{pmatrix} \quad Q = \frac{1}{7}\begin{pmatrix} 7 & 0 & 0 & -5 \\ 0 & 7 & 0 & -3 \\ 0 & 0 & 7 & -1 \\ 0 & 0 & 0 & 7 \end{pmatrix}$$

(iii)
$$P = \frac{1}{12}\begin{pmatrix} 5 & 3 & -1 & 0 \\ 4 & 0 & 4 & 0 \\ -7 & 3 & -1 & 0 \\ 1 & 0 & 0 & 1 \end{pmatrix} \quad Q = I_3$$

(iv)
$$P = \frac{1}{14}\begin{pmatrix} 0 & -4 & 2 & -4 \\ 0 & 2 & 6 & 2 \\ 0 & 3 & 2 & -4 \\ 1 & 1 & 0 & 1 \end{pmatrix} \quad Q = \frac{1}{14}\begin{pmatrix} 14 & 0 & 0 & -4 & -6 \\ 0 & 14 & 0 & 2 & -4 \\ 0 & 0 & 14 & -11 & -1 \\ 0 & 0 & 0 & 14 & 0 \\ 0 & 0 & 0 & 0 & 14 \end{pmatrix}$$

(P and Q are not uniquely determined.)

187

2. $\{(i), (iii)\}$, $\{(ii)\}$, $\{(iv), (v)\}$

Exercises 2.1

1. (i) 7,   (ii) 12,   (iii) −18,   (iv) −6,   (v) 0

2. (i) $(a-b)(a-c)(b-c)(a+b+c)$
   (ii) $(a-b)(a-c)(b-c)(a+b+c)(a^2+b^2+c^2)$
   (iii) $-2(a-b)(a-c)(b-c)(a+b+c)(a^2+b^2+c^2)$
   (iv) $8a(x-y)(x-z)(y-z)(x^2+y^2+z^2+a^2)$
   (v) $(a-b)(a-c)(b-c)(a+b+c)(a^2+b^2+c^2)$

3. $\theta = n\pi - \pi/12$ or $n\pi + 7\pi/12$ $(n = 0,1,2,\ldots)$

4. $0, b-c, (a+b+c)/2$

Exercises 2.2

1. (i) 5,   (ii) 11,   (iii) 2

2. $a, b, c, -(a+b+c)$

3. (i) 0,   (ii) ±4

5. $(-1)^{n^2-n}\, \alpha_{11}\alpha_{2n}\,\alpha_{3,n-1}\cdots\alpha_{nn}$

6. 2^{3k}

Exercises 2.3

1. (i)   is non-singular

2. $(\alpha-\beta)^2\,(\alpha-\gamma)^2\,(\beta-\gamma)^2\,(\alpha+\beta+\gamma)$

3. $\frac{1}{3}(-c)^r\,(8b^3-27c)$

5.
$$C = \begin{pmatrix} a & b & c \\ c & a & b \\ b & c & a \end{pmatrix}$$

6. (i) Yes   (ii) No

7. $(-5/3, -12/3, -16/3)$; $p=2, (10/3 + \mu, 2/3 + \mu, \mu)$; $p = -3, (5+\mu, 4+\mu, \mu)$
   μ arbitrary

8. (i) $k = -1, k = -4$,   (ii) $k \neq \pm 1, k \neq -4$,   (iii) $k = 1$

Exercises 2.4

1. (i)
$$\frac{1}{72} \begin{pmatrix} -3 & 5 & 9 \\ 18 & -6 & 18 \\ 6 & 14 & -18 \end{pmatrix}$$

 (ii)
$$-\frac{1}{9} \begin{pmatrix} 3 & -1 & -2 \\ 3 & -4 & 1 \\ -9 & -3 & 3 \end{pmatrix}$$

 (iii)
$$\frac{1}{3} \begin{pmatrix} 11 & -9 & 1 \\ -7 & 9 & -2 \\ 2 & -3 & 1 \end{pmatrix}$$

2. (i) $(1,0,0)$ (ii) $\dfrac{-1}{178} (41,17,169)$

3. $t = \pm 1$, $\dfrac{1}{(1-t^2)} \begin{pmatrix} 1 & -t & -t \\ -t & 1 & 1 \\ -t & t^2 & 1 \end{pmatrix}$

5. $\begin{pmatrix} x^2-x & -1 & x+1 \\ -2 & x^2-x-1 & 2(x+1) \\ -(x+1) & -(x+1) & (x+1)^2 \end{pmatrix}$, $x = \pm 1$, $\dfrac{1\pm\sqrt{3}i}{2}$

Exercises 3.3

1. (i) and (v)

2. (ii), (iii) and (iv)

3. (i), (iii) and (iv)

4. (i), (iii) and (vi)

5. (ii) and (iii)

189

Exercises 3.4

2. $(1,2,1), (0,1,-2)$

5. e.g. $\{(1,0,1,1), (1,0,2,4), (1,0,0,0), (0,1,0,0)\}$

7. e.g. $\{(1,1,0,-1), (4,-2,1,0), (1,0,0,0), (0,1,0,0)\}$

8. e.g. $\left\{ \begin{pmatrix} 1 & 0 \\ 0 & 0 \end{pmatrix}, \begin{pmatrix} 0 & 1 \\ -1 & 0 \end{pmatrix}, \begin{pmatrix} 0 & 0 \\ 0 & 1 \end{pmatrix} \right\}, \left\{ \begin{pmatrix} 1 & 0 \\ 0 & 0 \end{pmatrix}, \begin{pmatrix} 0 & 0 \\ 1 & 0 \end{pmatrix} \right\},$

$\left\{ \begin{pmatrix} 1 & 0 \\ 0 & 0 \end{pmatrix} \right\}, \left\{ \begin{pmatrix} 1 & 0 \\ 0 & 0 \end{pmatrix}, \begin{pmatrix} 0 & 1 \\ -1 & 0 \end{pmatrix}, \begin{pmatrix} 0 & 0 \\ 0 & 1 \end{pmatrix}, \begin{pmatrix} 0 & 0 \\ 1 & 0 \end{pmatrix} \right\}$

9. e.g. $\{(1,0,-1,0), (0,1,-1,0), (1,0,0,0), (0,0,1,-1)\}, \{(0,1,0,0)\}$

10. e.g. $\{(2,1,0,0), (-1,0,1,0), (3,0,0,1)\}$

12. $-4u_1 + 3u_2$

13. (i) Yes   (ii) No   (iii) Yes

14. u – No, v – Yes, e.g. $\{(2,-1,3,2), (-1,1,1,3), (3,-1,0,-1), (1,0,0,0)\}$
    $\{(1,0,4,-1), (2,-1,3,2), (1,0,0,0), (0,1,0,0)\}$

15. e.g. $\{(1,0,-1,1), (2,-1,0,1), (1,3,3,2)\}$, $S \cup \{y\}$

16. $4,8; 2,4$

17. $\{(0,1,1)\}$; $\{(1,1,0), (i,1+i,1)\}$

18. $\{1-t^2+t^3, t+t^2-t^3\}$, $\{1-t^2+t^3, 2+t-t^2+t^3, 1+t-t^2+t^3, t^2\}$

Exercises 4.1

1. (i) Yes   (ii) No   (iii) No   (iv) No

2. (i) Yes   (ii) Yes   (iii) Yes   (iv) No

3. (i) Yes   (ii) Yes   (iii) No   (iv) Yes   (v) Yes

Exercises 4.2

1. (i) $\begin{pmatrix} 1 & -1 & 0 \\ 1 & 2 & -1 \\ 2 & 1 & 1 \end{pmatrix}$ (ii) $\frac{1}{3}\begin{pmatrix} 5 & -11 & 12 \\ 2 & 1 & 0 \\ 2 & 4 & -6 \end{pmatrix}$

2. $\begin{pmatrix} 3 & 0 & 2 \\ 0 & -1 & 0 \\ -2 & 0 & -2 \end{pmatrix}$

4. If $S = \begin{pmatrix} s_{11} & s_{12} \\ s_{21} & s_{22} \end{pmatrix}$, matrices relative to $\{e_{11}, e_{12}, e_{21}, e_{22}\}$ are

(i) $\begin{pmatrix} s_{11} & 0 & s_{12} & 0 \\ 0 & s_{11} & 0 & s_{12} \\ s_{21} & 0 & s_{22} & 0 \\ 0 & s_{21} & 0 & s_{22} \end{pmatrix}$ (ii) $\begin{pmatrix} s_{11} & s_{21} & 0 & 0 \\ s_{12} & s_{22} & 0 & 0 \\ 0 & 0 & s_{11} & s_{21} \\ 0 & 0 & s_{12} & s_{22} \end{pmatrix}$

(iii) $\begin{pmatrix} 0 & -s_{21} & s_{12} & 0 \\ -s_{12} & (s_{11}-s_{22}) & 0 & s_{12} \\ s_{21} & 0 & (s_{22}-s_{11}) & -s_{21} \\ 0 & s_{21} & -s_{12} & 0 \end{pmatrix}$

5. $\begin{pmatrix} 1 & 1 \\ 2 & -1 \\ -1 & 0 \end{pmatrix}$, $\begin{pmatrix} 3 & -4 \\ -4 & 7 \\ 1 & -2 \end{pmatrix} = \begin{pmatrix} 1 & -1 & 0 \\ 0 & 1 & -1 \\ 0 & 0 & 1 \end{pmatrix}\begin{pmatrix} 1 & 1 \\ 2 & -1 \\ -1 & 0 \end{pmatrix}$, $\begin{pmatrix} -1 & 2 \\ 1 & -1 \end{pmatrix}$

6. $\begin{pmatrix} 3 & \lambda+2 & 1 \\ 1+\lambda+\mu & \lambda+\mu+\lambda\mu & \mu \\ 1+\lambda^2+\mu^2 & \lambda^2+\mu^2+\lambda\mu^2 & \mu^2 \end{pmatrix}$ (i) $\lambda \neq 1,\ \mu \neq 1,\ \lambda \neq \mu$

          (ii) $\lambda = \mu = 2$

7. $\{x^2, y^2, z^2, xy, xz, yz\}$

$\begin{pmatrix} \alpha^2 & 0 & 0 & 0 & 0 & 0 \\ 1 & \beta^2 & 0 & \beta & 0 & 0 \\ 1 & \gamma^2 & 0 & \gamma & 0 & 0 \\ 2\alpha & 0 & 0 & \alpha\beta & 0 & 0 \\ 2\alpha & 0 & 0 & \alpha\gamma & 0 & 0 \\ 1 & 2\beta\gamma & 0 & (\beta+\gamma) & 0 & 0 \end{pmatrix}$

8. $\{1, x, y, x^2, xy, y^2\}$, $\{1, x, x^2, x^3\}$

$$\begin{pmatrix} 1 & 0 & 2 & 0 & 0 & 4 \\ 0 & 1 & 0 & 0 & 2 & 0 \\ 0 & 0 & 0 & 0 & 0 & 0 \\ 0 & 0 & 0 & 1 & 0 & 0 \\ 0 & 0 & 0 & 0 & 0 & 0 \\ 0 & 0 & 0 & 0 & 0 & 0 \end{pmatrix}$$

Exercises 4.3

1. e.g. $\{(-1,2,3,0), (1,0,0,-1)\}$, $\{(1,1,0), (-1,2,3)\}$

2. 2,2

3.

If $P = \begin{pmatrix} 1 & 1 & 1 & 1 & . & . & . & 1 \\ 0 & 1 & 1 & 1 & . & . & . & 1 \\ 0 & 0 & 1 & 1 & . & . & . & 1 \\ 0 & 0 & 0 & 1 & . & . & . & 1 \\ . & . & . & & & . & . \\ . & . & . & & & . & . \\ . & . & . & & & . & . \\ & & & & & 1 & 1 \\ 0 & 0 & 0 & . & . & . & 0 & 1 \end{pmatrix}$

$Q = \begin{pmatrix} 1 & -1 & 1 & -1 & . & . & . & . & (-1)^n \\ 0 & 1 & -2 & 3 & . & . & . & . & (-1)^{n+1} \binom{n}{1} \\ 0 & 0 & 1 & -3 & . & . & . & . & (-1)^n \binom{n}{2} \\ 0 & 0 & 0 & 1 & . & . & . & . & (-1^{n+1} \binom{n}{3}) \\ . & . & . & . & & & & . . \\ . & . & . & . & & & & . . \\ . & . & . & . & & & & . \\ & & & & & & & -\binom{n}{1} \\ 0 & 0 & 0 & 0 & . & . & . & 1 \end{pmatrix} \in M_{n+1}(\mathbf{R})$

192

then

(a) (i)

$$A = \begin{pmatrix} 0 & 1 & 0 & . & . & . & 0 & 0 \\ 0 & 0 & 2 & . & . & . & 0 & 0 \\ 0 & 0 & 0 & & & & . & . \\ . & . & . & & & & . & . \\ . & . & . & & & & . & . \\ . & . & . & & & & . & . \\ 0 & 0 & 0 & . & . & . & 0 & n \\ 0 & 0 & 0 & & & & 0 & 0 \end{pmatrix}$$

(ii) $Q^{-1}AQ$ (iii) $P^{-1}AP$

(b) (i)

$$B = \begin{pmatrix} 1 & 1 & 1 & . & . & . & 1 \\ 0 & 1 & 2 & . & . & . & \binom{n}{1} \\ 0 & 0 & 1 & . & . & . & \binom{n}{2} \\ 0 & 0 & 0 & . & . & . & \binom{n}{3} \\ . & . & . & & & & . \\ . & . & . & & & & . \\ . & . & . & & & & . \\ 0 & 0 & 0 & . & . & . & 1 \end{pmatrix}$$

(ii) $Q^{-1}BQ$ (iii) $P^{-1}BP$

4. 2,2; $x = y = 0$

5. $\ker S = \{\alpha | \alpha \in \mathbf{R}\}$, $\mathrm{im}\, S = \left\{ \sum_{\substack{i,\,j=0 \\ 1 \leqslant i+j \leqslant n}}^{n} \alpha_{ij}\, x^i\, y^j | \alpha_{ij} \in \mathbf{R} \right\}$

$\ker T = \{\alpha + \beta x + \gamma y + \delta xy | \alpha, \beta, \gamma, \delta \in \mathbf{R}\}$

$\mathrm{im}\, T = \left\{ \sum_{\substack{i,\,j=0 \\ 2 \leqslant i+j \leqslant n}}^{n} \alpha_{ij}\, x^i\, y^j | \alpha_{ij} \in \mathbf{R}, \alpha_{11} = 0 \right\}$

$\{1, x, y, x^2, xy, y^2, \ldots, x^n, x^{n-1}y, \ldots, xy^{n-1}, y^n\}$

diag $(A_0, A_1, \ldots, A_n)$ where $A_i = $ diag $\underbrace{(i, i, \ldots, i)}_{(i+1) \text{ copies}}$ $(i = 0, 1, \ldots, n)$

diag $(B_0, B_1, \ldots, B_n)$ where $B_i = $ diag $(i^2 - i(2k+1) + 2k^2)$ $i = 0, 1, \ldots, n$;
$$k = 0, 1, \ldots, i$$

6. (i) $\mathbf{R}$ (ii) $\{f(x) | f^{(n)}(x) = 0\}$ $n \geqslant 1$ (iii) $\langle e^x \rangle$

7. (i) $\{0\}$, $M_2(\mathbf{R})$, (ii) $\{0\}$, $M_2(\mathbf{R})$, (iii) $\left\{ \begin{pmatrix} a & b \\ b & a+b \end{pmatrix} \middle| a, b \in \mathbf{R} \right\}$,
$\left\{ \begin{pmatrix} a & -b \\ a+b & -a \end{pmatrix} \middle| a, b \in \mathbf{R} \right\}$

8. $p - q = n - m$

9. 1

11. $V = V_2(\mathbf{R})$, $T(a,b) = (0, a)$, $a, b \in \mathbf{R}$. No

Exercises 4.4

1. $\lambda \neq 1/8$

2. $T(\alpha_1, \alpha_2, \alpha_3) = 1/3 (\alpha_1 + \alpha_2 + \alpha_3, 3\alpha_2 + 3\alpha_3, -\alpha_1 + 5\alpha_2 + 2\alpha_3)$

4. $\ker D \neq 0$

Exercises 4.5

2. (i) $a \neq 1, -14$ rank 4; $a = 1$ rank 2; $a = -14$ rank 3
   (ii) $a \neq 0, 1, 21/20$ rank 4; $a = 0$ rank 2; $a = 1$ rank 3; $a = 21/20$ rank 3

3. 2, 2

4. $t = 0, -1$; $t = 0$ rank 2; $t = -1$ rank 2

5. rank 3 unless $\beta = \gamma = 0$ when rank 1 or $\alpha = 0, \beta = \gamma \neq 0$ when rank 2

Exercises 5.1

1. (ii) and (iii)

2. $(0, 2, -1)$

3. (i) $\sqrt{6}$ (ii) $\sqrt{38}$ (iii) $\sqrt{2}$

4. (i) $\cos^{-1} 6/\sqrt{42}$, (ii) $\pi/2$

Exercises 5.2

1. (ii)

2. (i)

3. (i), (ii) and (iii)

4. None

5. (i) $3, 3, \sqrt{14}, -2, \cos^{-1} -2/9$

   (ii) $\sqrt{17}, \sqrt{46}, \sqrt{105}, 25, \cos^{-1} 25/\sqrt{782}$

   (iii) $1/\sqrt{3}, 1/\sqrt{2}, \sqrt{(5\pi^2-4)/6\pi^2}$, $-2/\pi^2, \cos^{-1} \dfrac{-2\sqrt{6}}{\pi^2}$

6. Sum of squares of diagonals equals sum of squares of sides

Exercises 5.3

2. (i) $<(1,1,0,0), (2,0,1,0), (-1,0,0,1)>$
   (ii) $<(1+i, 1-i)>$

3. $1-6x+6x^2$, $1/\sqrt{2}$, $6ac + 3ad + 3bc + 2bd = 0$

4. $\sqrt{2} \sin i\pi x$, $(i = 1, 2, \dots, n)$

5. (i) $\{(1,-1,1), (4,5,1), (2,-1,-3)\}$
   (ii) $\{(1,-1,1,1), (0,1,0,1), (3,1,-1,-1)\}$
   (iii) $\{(1,-1,i), (1+4i, 2-i, 5+i)\}$

195

6. (i) $\{(1/\sqrt{2},0,1/\sqrt{2}),(0,1,0),(1/\sqrt{2},0,-1/\sqrt{2})\}$

(ii) $\{1/2(1,i,1,i),1/2(i,1,i,1),1/\sqrt{2}(1,0,-1,0),1/\sqrt{2}(0,1,0,-1)\}$

7. $\{(1,0,0),(1,-1,0),(0,-1,1)\}$

8. $\{1,2x-1,6x^2-6x+1,20x^3-30x^2+12x-1\}$

9. (i) $\{e_1,e_2,e_3\}$; $(1,i,-i)$

(ii) $\{1,1/\sqrt{3}(2x-1),1/\sqrt{5}(6x^3-6x+1)\}$; $(4/3,\sqrt{3}/2,\sqrt{5}/6,(5/6,0,\sqrt{5}/6)$

10. $\{1,x,x^2-1/5,x^3-3/7x\}$

12. They are perpendicular

Exercises 6.2

1. (i)   (a) None   (b) None   (c) $1\pm\sqrt{2}i$

(ii)   (a) -1 (b) -1 (c) $-1,\pm i$

(iii) (a) 1 (b) $1,\pm\sqrt{2}$ (c) $1,\pm\sqrt{2},\pm i$

2. (i) $(1-x)(2-x)(3-x); 1,<(1,-1,0)>; 2,<(2,-1,-2)>;$
$3,<(1,-1,-2)>$

(ii) $(1-x)^3; 1,<(1,1,1)>$

(iii) $-x(x^2-2x-7); 0,(1,2+i,-1),-1$

(iv) $-(1-x)^2(1+x)(3-x); 1,<(1,0,0,-1),(0,1,-1,0)>;$
$-1,<(1,-1,-1,1)>,3,<(1,1,1,1)>$

(v) $(x-1)^2(x-2)(x-3); 1,<(1,-1,0,0),(1,0,-1,0)>$
$2,<(-2,4,1,2)>,3,<(0,3,1,2)>$

(vi) $(x-1)(x^2-2)(x^2+1); 1,<(1,0,1,-1,0)>;\sqrt{2},<(-1,\sqrt{2},1,0,0)>;$
$-\sqrt{2},<(1,\sqrt{2},-1,0,0)>;i,<(1,0,0,-1+i,1-i)>; -i,<(1,0,0,-1-i,1+i)>$

4. For $i=1,2,\ldots,n$, if X_i is an eigenvector corresponding to λ_i then X_i is an eigenvector corresponding to (i) $1/\lambda_i$ and (ii) λ_i^k $(k=1,2,\ldots)$

5. $0,<(1,0,\ldots,0)>$

196

7. Reflection: $1, < (\cos \phi, \sin \phi) >; -1, < (\sin \phi, -\cos \phi) >$
   Rotation: None

8. (i) $(-1)^{n-1} (\alpha_1 + \alpha_2 x + \ldots + \alpha_n x^{n-1} - x^n)$
   (ii) $(b-x)^{n-1} (1+a+ \ldots +a^{n-1} +b-x)$

Exercises 6.3

1. (i)
$$P = \begin{pmatrix} 0 & 1+i & 1-i \\ 1 & 1 & 1 \\ -1 & 1 & 1 \end{pmatrix}, P^{-1}AP = \text{diag}\,(-1,i,-i)$$

   (ii)
$$P = \begin{pmatrix} 1 & 2 & 1 \\ -1 & -1 & -1 \\ 0 & -2 & -2 \end{pmatrix}, P^{-1}AP = \text{diag}\,(1,2,3)$$

   (iii)
$$P = \begin{pmatrix} 4 & 4 & 2 \\ -1 & 0 & -1 \\ 0 & -1 & 0 \end{pmatrix}, P^{-1}AP = \text{diag}\,(0,1,2)$$

   (iv)
$$P = \begin{pmatrix} 1 & 0 & 1 \\ 0 & 1 & 0 \\ 1 & 0 & -1 \end{pmatrix}, P^{-1}AP = \text{diag}\,(2,1+i,2-2i)$$

   (v) Not diagonalizable

   (vi)
$$P = \begin{pmatrix} 1 & 0 & 1 & 2 \\ -1 & 0 & 1 & 2 \\ 1 & 1 & 1 & 2 \\ -1 & 1 & 1 & -1 \end{pmatrix}, P^{-1}AP = \text{diag}\,(0,1,2,-1)$$

   (vii) Not diagonalizable

2. $A: 1, < (3,5) >; 3, < (1,1) >; P = \begin{pmatrix} 3 & 1 \\ 5 & 1 \end{pmatrix}$

   $B: 2, < (1,-1) >$

3.
$$P = \begin{pmatrix} a^2 & 0 & a^2 & 1 \\ a^2 & a & -a & 1 \\ -1 & -1 & -a & 1 \\ -1 & 0 & 1 & 1 \end{pmatrix}, P^{-1}AP = \text{diag}\,(a,a,-a,1+a+a^2)$$

197

4.
$$\frac{1}{2}\begin{pmatrix} 2 & 2^{n+1}-2 & 3^{n+1}-2^{n+1}-1 \\ 0 & 2^{n+1} & 2.3^n-2^{n+1} \\ 0 & 0 & 2.3^n \end{pmatrix}$$

5.
$$P = \begin{pmatrix} 2 & 1 & -2 \\ 1 & 0 & 1 \\ 0 & -1 & 3 \end{pmatrix}, \quad X = P \begin{pmatrix} 1 & 0 & 0 \\ 0 & -1 & 0 \\ 0 & 0 & 0 \end{pmatrix} P^{-1} = \begin{pmatrix} 5 & -8 & 6 \\ 1 & -1 & 1 \\ -3 & 6 & -4 \end{pmatrix}$$

Exercises 6.4

1. (i) $x(x-2)$;   (ii) $(x-1)^2(x-2)$;   (iii) $(x-1)(x-2)$;   (iv) $(x-1)^2$

2. $\chi_A(x) = m_A(x) = x^n - a_{n-1} x^{n-1} - a_{n-2} x^{n-2} - \ldots - a_1 x - a_0$

3. $(x-1)(x+2)(x-3)$, $(x-1)(x+2)^2(x-3)$, $(x-1)^2(x+2)(x-3)$,
   $(x-1)^2(x+2)^2(x-3)$, $(x-1)^3(x+2)(x-3)$, $(x-1)^3(x+2)^2(x-3)$

5. (i) ,   (iii) diagonalizable

Exercises 6.5

1. (i)
$$\frac{1}{3}\begin{pmatrix} 2 & 2 & -1 \\ 2 & -1 & 2 \\ -1 & 2 & 2 \end{pmatrix}$$
(ii)
$$\frac{1}{7}\begin{pmatrix} 3 & 2 & 6 \\ -6 & 3 & 2 \\ -2 & -6 & 3 \end{pmatrix}$$

(iii)
$$\begin{pmatrix} 1/\sqrt{2} & 2/\sqrt{5} & 1/\sqrt{6} \\ 1/\sqrt{2} & 0 & -1/\sqrt{6} \\ 0 & 1/\sqrt{5} & -2/\sqrt{6} \end{pmatrix}$$
(iv)
$$\frac{1}{\sqrt{18}}\begin{pmatrix} 1 & 3 & 2\sqrt{2} \\ 1 & -3 & 2\sqrt{2} \\ 4 & 0 & -\sqrt{2} \end{pmatrix}$$

Exercises 6.6

1. (i) $3/2\, x^2 + 1/2\, y^2$;   (ii) $3/2\, x^2 + 1/2\, y^2$;   (iii) $2x^2 + 4y^2 + 5z^2$;
   (iv) $2x^2 - y^2 - z^2$;   (v) $7x^2 - 7y^2 + 14z^2$

2. (i) $1/\sqrt{2}\ (1,1),\ 1/\sqrt{2}\ (1,-1);\ x^2+y^2=4;$

   (ii) $1/\sqrt{13}\ (3,2),\ 1/\sqrt{13}\ (2,-3);\ x^2/6 - y^2/7 = 1;$

   (iii) $1/\sqrt{5}\ (1,2),\ 1/\sqrt{5}\ (2,-1);\ x^2/8 + y^2/3 = 1;$

   (iv) $1/\sqrt{2}\ (1,1),\ 1/\sqrt{2}\ (1,-1);\ x^2/4 + y^2/9 = 1;$

   (v) $1/\sqrt{2}\ (1,1),\ 1/\sqrt{2}\ (1,-1);\ (1/2,\ -5/2);\ x^2/4 - y^2/9 = 1;$

   (vi) $1/5\ (3,4),\ 1/5\ (-4,3);\ (7/5,\ -1/5);\ x^2/4 + y^2 = 1$

3. (i) $1/\sqrt{3}\ (1,1,1),\ 1/\sqrt{2}\ (1,-1,0),\ 1/\sqrt{6}\ (1,1,-2);\ 2x^2 - y^2 - z^2 = 1$

   (ii) $1/\sqrt{3}\ (1,1,1),\ 1/\sqrt{2}\ (1,0,-1),\ 1/\sqrt{6}\ (1,-2,1);\ y^2 + 3z^2 = 1$

   (iii) $1/\sqrt{6}\ (2,1,1),\ 1/\sqrt{3}\ (1,-1,1),\ 1/\sqrt{2}\ (0,1,1);\ x^2/8 + y^2/2 = 1$

   (iv) $1/\sqrt{2}\ (0,1,-1),\ 1/\sqrt{3}\ (1,-1,-1),\ 1/\sqrt{6}\ (2,1,1);\ x^2/7 + y^2/7 + z^2 = 1$

   (v) $1/\sqrt{2}\ (1,-1,0),\ 1/\sqrt{6}\ (1,1,-2),\ 1/\sqrt{3}\ (1,1,1);\ x^2 + y^2 - z^2 = 1$

   (vi) $1/11\ (7,6,6),\ 1/11\ (6,-9,2),\ 1/11\ (6,2,-9);\ 2x^2 + y^2 = 1$

   (vii) $1/3\ (1,-2,2),\ 1/3\ (-2,1,2),\ 1/3\ (2,2,1);\ 2z = 6y^2 - 3x^2$

   (viii) $1/\sqrt{3}\ (1,1,-1),\ 1/\sqrt{2}\ (0,1,1),\ 1/\sqrt{6}\ (2,-1,1);\ (-1/12,\ 1/8,\ 1/24);$
   $x^2 + 2y^2 - 3z^2 = -1/72$

199

Index

200